NETWORK OPTIMISATION PRACTICE:
A Computational Guide

NETWORK OPTIMISATION PRACTICE:
A Computational Guide

DAVID K. SMITH, M.A., Ph.D.

Department of Mathematical Statistics and Operational Research
University of Exeter

ELLIS HORWOOD LIMITED
Publishers · Chichester

Halsted Press: a division of
JOHN WILEY & SONS
New York · Brisbane · Chichester · Toronto

First published in 1982 by

ELLIS HORWOOD LIMITED
Market Cross House, Cooper Street, Chichester, West Sussex, PO19 1EB, England

The publisher's colophon is reproduced from James Gillison's drawing of the ancient Market Cross, Chichester.

Distributors:

Australia, New Zealand, South-east Asia:
Jacaranda-Wiley Ltd., Jacaranda Press,
JOHN WILEY & SONS INC.,
G.P.O. Box 859, Brisbane, Queensland 40001, Australia

Canada:
JOHN WILEY & SONS CANADA LIMITED
22 Worcester Road, Rexdale, Ontario, Canada.

Europe, Africa:
JOHN WILEY & SONS LIMITED
Baffins Lane, Chichester, West Sussex, England.

North and South America and the rest of the world:
Halsted Press: a division of
JOHN WILEY & SONS
605 Third Avenue, New York, N.Y. 10016, U.S.A.

© 1982 Christine and David Smith/Ellis Horwood Limited

British Library Cataloguing in Publication Data
Smith, David K.
Network optimisation practice: a computational guide.
1. Network analysis (Planning)
2. Mathematical optimization – Data processing
I. Title
003'028'54   QA402.3

Library of Congress Card No. 82-3028 AACR2

ISBN 0-85312-403-5 (Ellis Horwood Ltd., Publishers – Library Edn.)
ISBN 0-85312-445-0 (Ellis Horwood Ltd., Publishers – Student Edn.)
ISBN 0-470-27347-X (Halsted Press)

Typeset in Press Roman by Ellis Horwood Ltd.
Printed in the USA by the Maple Vail Book Manufacturing Group, New York

Table of Contents

". . . mister, is the Thames the water, or the hole it goes in?"
The policeman looked at her for a moment and then replied, "The water of course. You don't have a river without water."
"Oh," said Anna, "that's funny, that is, 'cos when it rains it ain't the Thames, but when it runs into the hole it is the Thames. Why is that, mister? Why?"

from *Mister God, this is Anna*
by Fynn (Collins 1974, used with the permission of the publisher)

Author's Preface

In the subject of network analysis, operational research rubs shoulders with one of the most elegant areas of mathematics: graph theory. It has been my experience, in both studying and teaching aspects of network analysis, that this is an area which is extremely appealing to the logical mind. For many of the ideas and methods which are described here, I hope that the reader's reaction will quickly be 'That's obvious'. But, obvious or not, the original discovery took time and effort, as does all worthwhile operational research and mathematics.

However, this is not just another book about network analysis. It is intended to be a source of useful ideas as well. The techniques outlined in the chapters which follow almost always involve a considerable amount of numerical computation. This is usually of a very straightforward kind, but its existence has two consequences for the student of the methods. The examples which can be given in the text of a book such as this will necessarily be small ones, otherwise the reader will be faced with seemingly endless pages of numerical computation. And the exercises which can be presented for student practice are also limited in size. However, most of the techniques can be implemented as computer programs with relative ease. It is to such programs that most users of the techniques will turn. Nonetheless some understanding of the underlying theory is important for users to make the fullest use of the information presented by the computer output.

In order that the gap between theory and practice can be bridged, this book presents listings of computer programs which demonstrate the algorithms. These programs have been written in two languages, BASIC and PASCAL. These languages are in wide use for teaching computer programming, and are widely available on computers of all sizes from desk-top machines to the largest of main-frame computers. The programs in the book are intended as an option; the book is self-contained without them, and it is good experience for the reader wishing to use a particular technique to write his/her own program for that, rather than rely on a ready-made one.

As an aid, either to the preparation of one's own programs, or to the understanding of those given here, there are notes on ideas for implementing the techniques as computer programs. These notes reflect my personal (and possibly idiosyncratic) approach to this problem. The notes should be taken as guidelines only; no two programmers will ever follow exactly the same mental processes in devising computer programs, and each will find deficiencies in the other's work.

There are thus two levels on which this book may be used; first as a text for students of network analysis, and second as a reference work for suggestions (and possible implementations) about the way in which the theory may be put into practice.

Much of the material in the book has been developed from courses which I have taught at Exeter. As a result, it is sometimes difficult to know where particular ideas and examples have come from, and so I apologise to any writers who may feel that their work has been reproduced by accident.

My thanks go to many people; to those students who have contributed by explaining their difficulties and their delights with the subject; to David Kohler who first introduced me to some of the material here; to John Beckett and Keith Tizzard for their encouragement; to James Gould for his work on the figures and to Sheila Westcott for typing the manuscript; and above all, to my wife Tina, for her encouragement, her forbearance, and her prayerful support.

David K. Smith
Exeter 1981

A note on the computer programs

The computer programs in the book have been developed over a considerable time on several different computers. However, for the purposes of the text, they have been transferred to two machines, and it is from listings prepared by these two that the printout here has been developed.

There were, at a recent count, over thirty dialects of BASIC available. Unhappily, not all of these are fully compatible with one another, so the task of producing fully transportable programs is a difficult one. The programs here have been written in a very basic form of BASIC, which should run without drastic changes on machines with any dialect. It is for this reason that readers who are familiar with any of the more sophisticated BASIC compilers or interpreters may find constructions which appear archaic. For instance, all the lines have numbers; there is only one statement on each line; logical statements are all of the form: IF simple expression THEN line number, or IF simple expression GOSUB line number. Obviously, then, there is scope for revision of the programs if an advanced dialect of BASIC is available — but I hope that the ideas presented will help in the production of such revision.

PASCAL compilers tend to keep to the standard established by Jensen and Wirth [6]. The programs are therefore more readily transportable between computers, although there may be some implementation-defined parameters which need to be amended, such as the size of sets of integers, and (where it is used) the constant inf. This has been used as a suitable large number in place of infinity; it should not exceed (maxint-1)/2, or else errors are likely.

All the programs have been designed to run interactively, using a teletype or visual display unit of eighty characters width. Minor modifications may be needed for narrower displays, mainly in the output of results. If a batch mode of running is to be used, then the prompts and comments on input should be removed, and a schema for data input drawn up.

The algorithms have all been presented as programs. Many of the methods could be presented as a subroutine, or set of linked subroutines. This approach was considered and rejected, so that each could be demonstrated in a self-contained form, with input and output, and comments which are appropriate for the particular program. It also, in many cases, allows for development of the algorithm by the user, without constant reference to a host of other sub-routines. It will be seen that verification of the data takes up a significant part of most of the programs. This is deliberate. Although not every error will be picked up by the checks which are made, it is hoped that many will be found, and corrected without the loss of all the data. Few things are more infuriating than to find that a typing error on the last line of data input has wrecked the computer run!

   To facilitate reproduction of the programs, they have been processed so as to fit the page size of this book. This has meant that some lines have been broken into two parts. Such lines may be identified readily, as follows. In PASCAL programs, the second part of a broken line starts with the symbol "$\neq\neq$" followed by three spaces; in BASIC programs, any unnumbered line in the listing is the second part of a broken program line. In each case, the breaks have been made immediately after a space, a comma or a semi-colon from the original program.

   Any of the programs here may be freely used for educational purposes; suitable acknowledgement should be given.

   Presenting a book of computer programs for others to use is a little like putting one's head on the block. There is bound to be some feature which does not appeal to one reader or another, and which will lead to disappointment. If you are one such reader, my sincere apologies; but, as an old saying goes, 'There is more than one way to skin a cat'! If you have suggestions to make about any of the programs, please write to me with them — and the same goes for any errors, of which I hope there are none!

David K. Smith

1

Introduction

This book is about networks. It is sensible, therefore, to start with some ideas and definitions about this subject, and something of their value and use. If one were to consult a dictionary, one formal definition would be:

> **network** any structure in the form of a net: a system of lines, e.g. railway lines, resembling a net: system of units, as e.g. buildings, agencies, groups of persons, constituting a widely spread organis-ation and having a common purpose: an arrangement of electrical components: a system of stations connected for broadcasting the same programme (radio and TV).

(*Chambers 20th Century Dictionary* (1977), used with the permission of the publisher)

This is not a book about structures; nor is it about organisations, electricity or the broadcast media. The kind of networks to be studied will fall into the part of the definition which is 'a system of lines. . . resembling a net', which means that the lines will be joined together in some way.

Not only will the lines be joined together in groups, but the lines will have various properties associated with them, and in some cases, the points of union of two or more lines will also be endowed with properties of interest. The whole system will normally be a convenient representation of some system in the world. There are many everyday systems and phenomena which can be readily recognised as networks in their own right, since they satisfy the definition above; the circuit diagram for any electrical appliance is one such, and road/rail maps give other familiar examples. Yet other situations in day-to-day life give rise to networks, though in less obvious ways. The priority for a range of jobs in a factory and the possible routes for promotion in a large organisation are such. Frequently, a network is used to convey information in a concise form, as in Fig. 1.1.

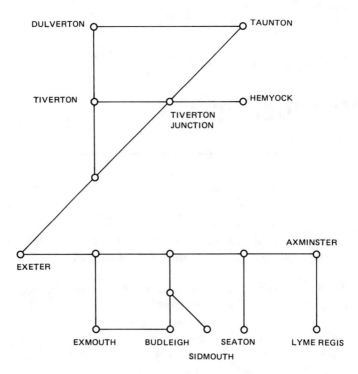

Fig. 1.1 – The railways of east Devon (circa 1905), (not to scale).

The study of networks in operational research and mathematics is devoted to finding attributes of them which satisfy certain conditions, using the properties of the lines (usually known as arcs) and the arrangement of these. In one chapter, the property associated with a line is the length between its two ends; the ends are places where it is possible to move from one line to another, and the condition of interest is the set of lines which join two points together to give the shortest possible distance between them. In other chapters, the property associated with the lines is the capacity for flow of some commodity; the aim then will be to find ways of sending this commodity through the network without violating the limits on the amount sent along any one line. Later, other properties of the parts of networks will be introduced, and other objectives studied.

It is necessary to introduce several formal definitions which will be used throughout the book. These give some structure to the informal ideas which the definition of a network has conjured up. Such formal statements will not be essential in the later discussion – through it is important that their existence and whereabouts are known!

1.1 FUNDAMENTAL DEFINITIONS

A **graph** (N,A) is a set of points N, and a set A of ordered pairs of these points. The members of N are known as **nodes** and N is known as the **node set**. The members of A are known as **arcs** with A as the **arc set**. In many cases, the ordering of the pairs of A is important, and then the arcs are known as **directed arcs**; usually the context will indicate whether directed arcs are being used. An arc (i,j) is a link from i to j, and i is known as either the **start node** or the **tail node** of the arc; j is known as the **end node** or the **finish node** or the **head node**.

It is usual to depict the graph (N,A) using lines for arcs and small circles for nodes; the direction of arcs can then be indicated with an arrow. Where there are arcs in pairs (i,j) and (j,i), then this pair of arcs can either be represented by two lines, each with arrows, or by one without any indication of its direction. Typical graphs are shown in Figs. 1.2 and 1.3.

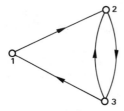   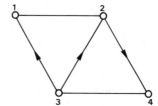

Fig. 1.2.   Fig. 1.3

Figure 1.2 may be described as: $\left(\underset{N}{\underbrace{\{1,2,3\}}}, \underset{A}{\underbrace{\{(1,2), (2,3), (3,2), (3,1)\}}}\right)$

and Fig. 1.3 as: $(\{1,2,3,4\}, \{(1,2), (2,1), (2,4), (3,1), (3,2), (3,4), (4,3)\})$.

The position of the arcs and nodes in a diagram is not normally important, although it is desirable to produce as tidy a figure as possible. Figure 1.4 represents the same graph as Fig. 1.3.

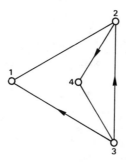

Fig. 1.4.

If all the arcs in a graph occur in pairs, so that if the arc (i,j) is present, so is arc (j,i), then the graph is called an **undirected graph** and the arcs are sometimes referred to as **edges**.

If there are some arcs which occur in pairs, and some which do not, then the graph is sometimes known as a **mixed** graph.

A **network** is a graph in which all the arcs possess some property or properties in addition to their two end nodes. Numerous properties are possible, and in this book, study will be made of some of the most commonly found properties of the arcs in a network. Since there may be several properties associated with each arc in a network, there is no universal representation of a network in a conventional mathematical form. The set of nodes and arcs which define the graph of the network are usually referred to as (N,A), and the arc properties P may in some cases be added to form a triple (N,A,P), or referred to separately.

In addition to the properties of individual arcs, algorithms for networks make extensive use of the properties of arcs taken in groups or sets. These are usually groups which are connected to one another at nodes. Four such sets are found frequently.

A **chain** between two nodes s and t is a set of arcs which join these two nodes; here, 'joining' is taken in the sense that there is one arc connecting s to an intermediate node i_1, an arc connecting i_1 to a node i_2, and so on, until there is the rth intermediate node i_r, which is connected by an arc to node t. There will then be $r + 1$ arcs in the chain. The arcs do not all need to be in the same sense; some may be 'from s to t', the others 'from t to s'.

A **path** from s to t in a network is a chain whose arcs are all in the sense 'from s to t'.

A **cycle** is a chain whose starting node is the same as its finishing node; a **circuit** is a path with this characteristic. In some cases it is desirable, for chains, to distinguish between the arcs which are in the sense 'from s to t', which are known as **forward arcs** and the others which are known as **reverse arcs**.

There is considerable latitude in these definitions, and it is sometimes found necessary to restrict them so that cycles within chains and paths do not occur. If this restriction is imposed, then the adjective **simple** is applied to the resultant chains and paths.

These ideas are illustrated in Fig. 1.5.

The set of arcs $(1,2)$, $(2,4)$, $(4,3)$, $(3,5)$, $(4,5)$, $(4,6)$, $(7,6)$ is a chain from node 1 to node 7. It is not a simple chain because there is a cycle $((4,3), (3,5), (4,5))$. It is not a path because there are two reverse arcs $((4,5)$ and $(7,6))$.

The set $(1,2)$, $(2,4)$, $(4,5)$, $(5,6)$ is a path from node 1 to node 5. This is a simple path, since there are no cycles in it. $(1,3)$, $(3,5)$, $(4,5)$, $(4,1)$ is a simple cycle, but not a circuit, as the arc $(4,5)$ is in the opposite sense to all the rest.

$(1,3)$, $(3,2)$, $(2,4)$, $(4,3)$, $(3,5)$, $(5,6)$, $(6,4)$, $(4,1)$ is a circuit, but the presence of the cycle means it is not a simple one.

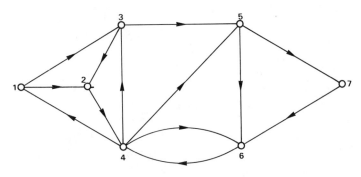

Fig. 1.5.

1.2 PARTIAL GRAPHS AND SUB-GRAPHS

Given a graph (N,A) it is possible to consider the graphs which result from taking subsets of A, or of N and A. The graph which results from reducing the set of arcs to a set A', while retaining the same set of nodes, is (N,A') and is termed a **partial graph** of (N,A). If the set of nodes is reduced to N', and the set of arcs reduced to include all the arcs both of whose nodes are in N', a **subgraph** of (N,A) is formed. (Thus, the subgraph (N',A') of (N,A) has $N' \subset N$ and $A' = \{(i,j) : (i,j) \in A, i,j \in N'\}$). The two definitions may be combined to give a **partial subgraph**.

Graphs may be connected or disconnected. A **connected** graph is one in which every pair of nodes is connected by a chain. In **disconnected** graphs, there are pairs of nodes which cannot be thus linked. In Fig. 1.6, there is a disconnected graph with three separate parts or **components**.

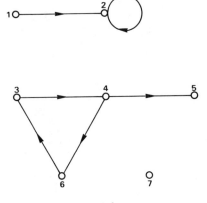

Fig. 1.6.

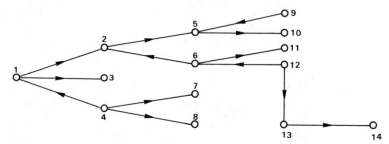

Fig. 1.7 – A typical tree.

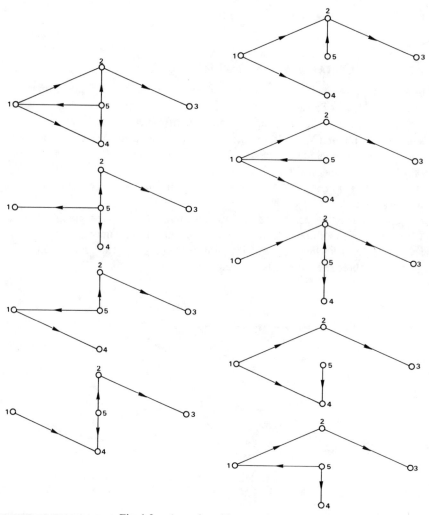

Fig. 1.8 – A graph and its spanning trees.

One type of graph which has particular properties of interest in optimisation of graphs and networks is a **tree**. A tree is a connected graph without any cycles: in Fig. 1.7, a typical tree is shown. There are numerous equivalent definitions of a tree, such as:

a tree is a connected graph with n nodes and $n-1$ arcs;

a tree is a graph with every pair of nodes linked by one and only one chain;

a tree is a connected graph, such that the removal of any of its arcs would yield a disconnected graph.

Trees form partial graphs of other graphs, and in such cases are known as **spanning trees**. There are usually several spanning trees for any graph, and in Fig. 1.8, a graph is shown with all its spanning trees.

1.3 THE MATRIX REPRESENTATION OF GRAPHS

It is difficult to represent graphs diagrammatically for numerical (and especially computational) purposes. It is therefore more usual to use some numerical structure for this, and matrices have been found to be the most convenient form for such purposes. The information which must be conveyed is the position of the arcs in terms of the nodes which they connect, and there are two widely used ways of conveying this in a matrix. The first is known as an **adjacency matrix**, and is a square matrix, with one row and one column for each node. The numerical entries are either zero or one. Where there is an arc (i,j) in the graph, then the entry in the ith row and jth column is 1; if there is no such arc, then a zero is placed in this position. An alternative to this is the **node-arc incidence matrix** (also known as the incidence matrix) which has one row for each node and one column for each arc. This is made up of entries $+1,0$ and -1. If the kth arc is (i,j), then the kth column of this matrix has $+1$ in the ith row, -1 in the jth row, and zeroes everywhere else. If the kth arc is (i,i) (that is, a loop from i to itself) then the kth column is wholly zero. Examples of a graph and its matrices are shown in Fig. 1.9.

Each matrix has its advantages and disadvantages. The adjacency matrix records information about every arc except where there are parallel arcs, while the incidence matrix may contain columns which are ambiguous as a result of the presence of loops. On the other hand, the incidence matrix has one column for each arc, and so is more useful when properties of arcs are being considered. (There is a lack of agreement amongst writers about the names, and the structure, of these matrices; some refer to the node-arc incidence matrix as 'the matrix of the graph', and may choose not to include any loops at all, so that all columns of this matrix have one entry of $+1$ and one of -1. Others extend the concept of the node-arc incidence matrix to undirected graphs, with a column for each undirected arc, which is wholly zero except for an entry of $+1$ in the rows corresponding to the ends of the arc.)

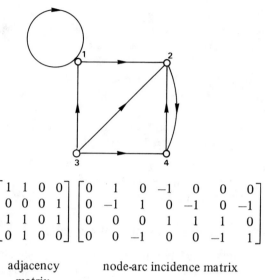

$$\begin{bmatrix} 1 & 1 & 0 & 0 \\ 0 & 0 & 0 & 1 \\ 1 & 1 & 0 & 1 \\ 0 & 1 & 0 & 0 \end{bmatrix} \begin{bmatrix} 0 & 1 & 0 & -1 & 0 & 0 & 0 \\ 0 & -1 & 1 & 0 & -1 & 0 & -1 \\ 0 & 0 & 0 & 1 & 1 & 1 & 0 \\ 0 & 0 & -1 & 0 & 0 & -1 & 1 \end{bmatrix}$$

adjacency   node-arc incidence matrix
matrix

Fig. 1.9 – A graph and its matrix representations.

Properties of networks are normally stored in the form of matrices similar to the adjacency matrix. Thus, a commonly encountered property of an arc is its length; the lengths of all the arcs in a network could be stored in a square matrix, where the entry in the ith row and the jth column was the length of the arc between node i and node j, with ∞ in those cases where there was no arc (i,j). For large networks, computational efficiency may mean that the best way to store the properties of the arcs is by means of a list, with three entries for each arc (start node, finish node, value of the property).

This book is concerned with optimisation methods for networks and graphs. In each chapter, different aspects of such optimisation will be described. The results may be optimal values for some property associated with the arcs of the network, or optimal sets of arcs, or optimal paths and chains. Reference will be made to many of the introductory ideas described above, but apart from this, the chapters can be studied in any order (with a few cross-references where these are absolutely essential).

2

Simple Algorithms for Graphs

The first algorithms to be considered in this book are those concerned with the most elementary of graph properties. These are ones which can be easily performed by hand, but it is useful to study the way that they can be implemented as computer programs.

The chapter is divided into two parts; the first develops the concept of trees from their definition, given in the preceding chapter, and develops two algorithms for finding a minimal spanning tree in a given network. The second part is related to this, and deals with algorithms for connectedness. If a network is connected, then it will have a spanning tree; if there are several components, then there can be no spanning tree. Problems relating to connectedness and spanning trees arise in practice from studies of communication networks, where it is wished to find whether a set of points are connected, and if they are, what the minimal set of arcs for linking them is. In addition, trees have many everyday uses, which are described briefly below.

2.1 TREES

Many human events can be represented by a tree, and the simplest such is a family tree (from which the term is derived). Family trees may be drawn in one of several ways, such as by taking one person and representing each of his/her ancestors by a node, with arcs representing parenthood. Alternatively, a couple may be represented by a node, and their descendants (together with their spouses) by further nodes, with arcs to indicate the link of parents to child. Strictly speaking, if such family trees are extended over several generations, the result need not be a tree in the sense used in graph theory. Loops are possible, since one individual may be a parent of two or more members of a family, whose descendants later intermarry.

Trees are also appropriate for representing the way that articles are sorted. Letters and parcels in the post are sorted according to their destination, and this process is carried out in a number of stages, some of which have been mechanised in many countries. The use of postal codes (under the name of 'postcode' or 'zipcode') is a step towards clearly defined sorting procedures. In Fig. 2.1, the stages in sorting a letter whose destination has the (British) postcode EX4 4PU are shown.

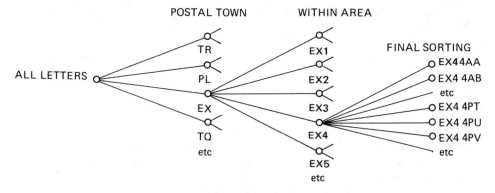

Fig. 2.1 — Sorting a letter addressed to EX4 4PU.

Other trees could be used to represent the way that a child sorts building bricks (by colour and by size), or the way that exhibits are sorted for an art exhibition (into those which are to be displayed, and the others; and then into arrangements in the rooms of the gallery).

Within a given graph (N,A), there are many partial subgraphs which are trees. A spanning tree for the graph is a set of arcs which forms a tree and a partial graph. So, a spanning tree is a tree which contains a path between every pair of nodes in the graph, and is, in a sense, 'minimal', in that every arc is needed. Removal of any one arc from a spanning tree means that the graph is no longer 'spanned' in the sense of there being a path between all pairs. The addition of an arc to a spanning tree results in a graph which is not a tree, and which has loops giving alternative paths between some pairs of nodes. In Fig. 2.2, a graph (N,A) is shown, together with one of its spanning trees. Figure 2.3 shows the result when some of the constituent arcs of this tree are removed one at a time, and it will be seen that in each case the result is no longer a partial graph. Figure 2.4 shows the same spanning tree with other members of (N,A) added, and the results in each case are not trees.

When there is a distance measure D associated with the arcs, to give a network (N,A,D), it is possible to define the length of a spanning tree. This is defined as the sum of the lengths of the arcs which constitute the tree, and, clearly, different trees may be compared in respect of their lengths. (A distance measure need not refer to a physical length, whether in metres or miles. It may refer to a cost, or to a fuel consumption, or to a travel time. The key idea of such a measure

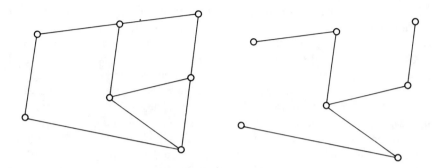

Fig. 2.2 – An undirected graph, and one of its spanning trees.

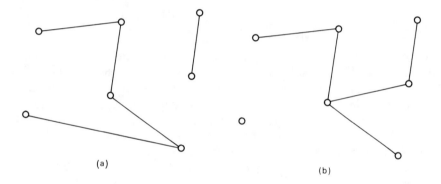

(a)   (b)

Fig. 2.3 – The effect of deleting arcs from the spanning tree of Fig. 2.2.

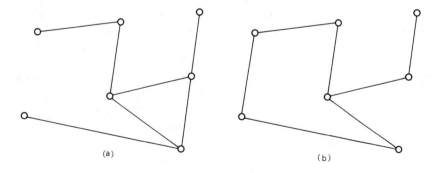

(a)   (b)

Fig. 2.4 – The effect of adding arcs to the spanning tree of Fig. 2.2.

is that the value of the property associated with two or more arcs is the sum of the properties for the individual arcs.) One spanning tree of especial interest is the minimal spanning tree; this is the spanning tree whose length is least, for a given network. Several trees may have the same length. The value of a minimal spanning tree in a network is that it may be used in problems of resource alloca-tion; if one is interested in the set (from a network of possible links between nodes) of roads, power supply cables, or gas pipelines which connect a number of localities with the smallest total distance or cost, then the minimal spanning tree will yield this set. (It is assumed in such a problem that the only places where there may be connections between links are the localities; if this is not so, then a much more difficult problem results, since the positions where arcs meet are variables to be included in the solution of the problem. A detailed discussion of this 'Steiner problem' lies outside the scope of this book, but interested readers will find details in Gilbert & Pollack [10].)

2.2 THE MINIMAL SPANNING TREE

Construction of a minimal spanning tree is straightforward. It is evident that the arcs in such a tree will be taken from the shortest arcs in the network, in order that a minimal total be achieved. An iterative procedure can be used to construct the tree, starting with the shortest arc and expanding the set of arcs which are in the tree gradually. The set will initially comprise one arc, whose length is less than or equal to the length of every other arc. To this set arcs will be added chosen so as:

(a) to be the shortest available,

and (b) to form a tree or set of trees with the arcs which have already been put into the set.

There are two main ways of achieving this iteration. In one, all the arcs are ranked by length, and are added to the set if they do not form a loop with those arcs which are already in the set; this method is due to J. B. Kruskal [11], and may lead to several sub-trees being grown simultaneously, and then being joined together. The second method, due to R. C. Prim [12] is sometimes called the 'greedy algorithm'. The spanning tree is grown steadily, from one initial arc; at each stage of the algorithm, the arc which is added is the shortest arc remaining which has one vertex in the tree. So, to select an arc, those which are not yet in the partial tree are scanned, and divided into two sets, corresponding to those which have one (and only one) node in the tree, and those which have none or two. Those with two would form loops, if they were to be added, and those with no nodes could not be added since they would not form a tree. The latter set is discarded, and the shortest arc from the former (chosen arbitrarily in cases where there is a tie) is added to the partial tree. This adds one further node to the tree, so the sets have to be altered, and the process repeated until the partial tree is a spanning tree for the network. In terms of the two sets, this corresponds to the first set being empty, or equivalently, the second set being wholly made up of arcs which have two nodes contained in the partial tree. Where Kruskal's method may create several small trees before the spanning tree is complete, Prim's method steadily expands one partial tree to become a spanning tree. These may be stated formally as:

2.2.1 Kruskal's algorithm for a network (N, A, D)
step 0   Initialise the graph T with n nodes and no arcs.

step 1   Create a list L of arcs from N in ascending order of length. (Rank arbitrarily those arcs with the same length.)

step 2   Select the arc (i, j) from the head of L. If it forms a circuit in T, delete it from L, and repeat step 2. Otherwise transfer it from L to T.

step 3   If T is a tree, stop; otherwise repeat step 2.

2.2.2 Prim's algorithm for a network (N, A, D)
step 0   Initialise the graph T to have one node, selected arbitrarily, and no arcs.

step 1   Select the arc (i, j) whose length is least, from among those arcs (i, l) which have i in T, l not in T. Add this arc to T, and add j to the node set of T.

step 2   If T is a spanning tree for the graph (N, A), stop; otherwise, repeat step 1.

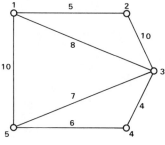

Fig. 2.5.

2.2.3 Worked Example
Consider the network shown in Fig. 2.5. Here, the node set N is $\{1,2,3,4,5\}$, the arc set A is $\{(1,2), (2,3), (3,4), (4,5), (1,5), (1,3), (3,5)\}$, and the distance matrix D is

$$
\begin{bmatrix}
\infty & 5 & 8 & \infty & 10 \\
5 & \infty & 10 & \infty & \infty \\
8 & 10 & \infty & 4 & 7 \\
\infty & \infty & 4 & \infty & 6 \\
10 & \infty & 7 & 6 & \infty
\end{bmatrix}.
$$

Applying Kruskal's method to this, the tree T grows as follows:

step 0 $T = (\{1,2,3,4,5\}, \emptyset)$.

step 1   The ordered list L is (3,4), (1,2), (4,5), (3,5), (1,3), (1,5), (2,3).

step 2   Arc (3,4) is transferred from L to T,
         L is (1,2), (4,5), (3,5), (1,3), (1,5), (2,3),
         T is $(\{1,2,3,4,5\}, \{(3,4)\})$.

step 3 T is not a tree, so repeat step 2.

step 2   Arc (1,2) is transferred from L to T,
         L is (4,5), (3,5), (1,3), (1,5), (2,3),
         T is $(\{1,2,3,4,5\}, \{(3,4), (1,2)\})$.

step 3 T is not a tree, so repeat step 2.

step 2   Arc (4,5) is transferred from L to T,
         L is (3,5), (1,3), (1,5), (2,3),
         T is $(\{1,2,3,4,5\}, \{(3,4), (1,2), (4,5)\})$.

step 3 T is not a tree, so repeat step 2.

step 2   Arc (3,5) is now at the head of the list L. If this arc were to be added
         to T, it would form a circuit with those arcs which already belong to T
         (in this case with the arcs (3,4), (4,5)). So delete it from L, leaving L as
         (1,3), (1,5), (2,3), and repeat step 2.

step 2   Arc (1,3) is transferred from L to T,
         L is now (1,5), (2,3),
         T is now $(\{1,2,3,4,5\}, \{(3,4), (1,2), (4,5), (1,3)\})$.
         T is now a spanning tree, so stop. T is shown in Fig. 2.6, and has total
         length 23.

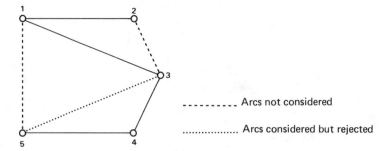

Fig. 2.6.

Using Prim's algorithm for the same network, the tree grows in a different manner, although the result is still the same.

step 0   Let T be initialised as $(\{1\}, \emptyset)$.

step 1   Consider arcs (1,2), (1,3), (1,5). Of these, (1,2) is the shortest,
         so is added to T; node 2 is added to the node set of T, yielding
         $T = (\{1,2\}, \{(1,2)\})$.

step 2 T is not a spanning tree, so repeat step 1.

step 1   Consider arcs $(1,3)$, $(1,5)$, $(2,3)$. Of these, $(1,3)$ is the shortest, so is added to T; node 3 is added to the node set of T, yielding $T = (\{1,2,3\}, \{(1,2), (1,3)\})$.

step 2 T is not a spanning tree, so repeat step 1.

step 1   Consider arcs $(1,5)$, $(3,4)$, $(3,5)$. (Arc $(2,3)$ is not considered since both its ends are in the node set of T.) Of these, $(3,4)$ is the shortest, so is added to T; node 4 is added to the node set of T, yielding $T = (\{1,2,3,4\}, \{(1,2), (1,3), (3,4)\})$.

step 2 T is not a spanning tree, so repeat step 1.

step 1   Consider arcs $(1,5)$, $(3,5)$, $(4,5)$. Of these, $(4,5)$, is the shortest, so is added to T; node 5 is added to the node set of T, yielding $T = (\{1,2,3,4,5\}, \{(1,2), (1,3), (3,4), (4,5)\})$.

step 2 T is now a spanning tree, and is the minimal spanning tree for the network.

2.2.4 Notes

(a) The minimal spanning tree need not be unique. The spanning trees which these algorithms produce depend on the manner in which the possible arbitrary decisions in each have been taken. In Kruskal's method, the arcs of equal length were ranked in an arbitrary order. For some networks, a change of order would have led to another spanning tree T being produced. In Prim's method, the first node was chosen arbitrarily. Had another node been chosen, in some networks, another spanning tree would have been found. (In the case used as an example, the minimal spanning tree is unique; in exercise (2.1) at the end of the chapter, a network is given which has several minimal spanning trees.)

(b) The two algorithms have generated the **minimal** spanning tree, in the sense of the tree with least total length. In an analogous fashion, it is possible to define a **maximal** spanning tree, which would be the spanning tree with greatest total length. No new algorithm is needed to find this, since either of the two described may be used. Suppose the maximal spanning tree in (N, A, D) is desired. Create a network (N, A, D') in which $d'(i,j) = M - d(i,j)$, where M is some large number chosen to be greater than the length of every link in (N,A,D). Then the arcs in the minimal spanning tree of (N,A,D') are the arcs of the maximal spanning tree of (N,A,D).

(c) Neither of the algorithms uses the length of the tree as a criterion for construction of the tree; only the lengths of arcs are used, and these are ranked into an order to identify the next arc to be added to the tree. It is not necessary to use the length of arcs as the criterion, and the ranking of arcs could be on the basis of some other measure, or on the basis of the value of some function; it is necessary that the overall objective is to minimise the sum of the measures assigned to the arcs. So, one could seek the spanning tree for which the sum of the logarithms of the arc lengths was the minimum, and this would correspond to minimising the product of the lengths of the arcs in the tree. A moment's reflection should show that this spanning tree is the same as the minimal spanning tree, since the ordered list of arcs will be the same in each case. Additionally, the minimal spanning tree found by Kruskal's algorithm and by Prim's algorithm is optimal for the problem 'Find the spanning tree whose longest arc is as short as possible' (exercise (2.3)).

2.2.5 Implementation

For each of the algorithms, it is necessary to devise a convenient representation of the three parts of the network's specification, the set of nodes, the set of arcs and the distances along these arcs. There are several possible methods for this. The most convenient assume that the nodes are numbered consecutively from 1 to the total number of nodes, so that the set of nodes is simply the set of integers $\{1, 2, 3, \ldots, n\}$. Then the arcs, and their lengths, may be represented in one of two straightforward fashions. First, they may be stored as three one dimensional arrays (vectors) representing the start node, finish node and length. The size of these arrays will be the number of arcs in the network. Alternatively, the matrix of arc lengths may be created; this will have entries only where the corresponding arcs exist, with infinity for non-existent links. (In practice, as infinity is a difficult concept for inclusion in programs, a large number is used instead.) Each of these schemes has its own benefits, and its own problems. In some cases, the choice between them has to depend on the size of problem being studied. The programs which follow have used both schemes.

As well as there being a choice of ways of representing the network, there is a wide choice of ways of inputting the network into computer memory. Some checks are essential, if spurious data are to be eliminated. In the programs here, it is assumed that each arc will be input separately. For a very large network, with few infinite entries in the distance matrix, then it would be more convenient to directly input this matrix, and suitably modify the data handling routines.

The sections which follow deal with specific matters concerning each of the algorithms.

Implementing Kruskal's algorithm

From a programmer's point of view, the most difficult part of Kruskal's method is the test for the formation of a loop. If an arc (i,j) is being considered for inclusion in the tree, and either i or j or both are not yet in it, then a loop cannot be formed by including the arc. But, if both i and j are in the partial tree, then it is possible that adding this arc will lead to a loop being formed. A loop will only be formed if there is a chain between i and j in the partial tree.

In order to find whether there is a chain between i and j, the BASIC program gives a label to all the nodes in the partial tree which are connected to node i. The set of labels is built up by repeatedly scanning through the arcs of the tree, to find those which have one end labelled and one not. The unlabelled node can then be labelled, and the list scanned again. If a scan through the arcs fails to label any more nodes, then the set of labelled nodes is the set of all nodes which are connected to i; if j is labelled, then there is a chain from i to j, and the arc (i,j) would form a loop. If not, then (i,j) can be safely added to the partial tree.

In the PASCAL program, a different approach is made. Here a list is made of the nodes which are connected to i, listed in the order of being found by the program. This list is built up, by listing all the nodes which are connected to i directly, and then examining these nodes and adding those nodes which are connected to them. So, there are two pointers to the list, one for the node being examined, and one for the foot of the list; only nodes which are not on the list can be added, and these are identified using a set which matches the list. The whole scanning process is implemented as a Boolean function, which is true if there would be a loop formed by the inclusion of a particular new arc.

The tree is stored by a vector corresponding to the arcs; arcs which are in the spanning tree are given a label which is sufficient to identify them when the tree is being output. The programs do not wholly follow step 1 of the algorithm, as the list L is not ordered completely. Instead, the arcs are gradually sorted and as successively larger arcs are placed in order, each is tested for inclusion in the tree.

```
program prog2p1(input,output);
  const maxarcs = 30;
        maxnodes = 30;
  type nodenos  = 0..maxnodes;
       nodeset = set of nodenos;
       arcarr = array[1..maxarcs] of integer;
  var  i,j,d : arcarr;
       st,t8,t9,arcs,n : integer;
       s : nodeset;

function count(n:integer; var
thisset:nodeset): integer;
{ count takes as value the number of
  members of set thisset }
var i,j : nodenos;
begin
  j := 1;
  for i := 1 to n do if (i in thisset)
    then j := j+1;
    count := j-1
end;

function least(n:integer; var
thisset:nodeset): integer;
{ least takes the value of the
  numerically least member of thisset }
var j : nodenos;
begin
  if (count(n,thisset))>0) then
    begin
      j := 1;
```

```
        while (not(j in thisset)) do
          j := j+1;
          least := j
      end
      else least := 0
    end;

function last(n:integer; thisset: nodeset)
  : nodenos;
{ last is the largest member of thisset }
var j: nodenos;
begin
  if (count(n,thisset))>0) then
    begin
      j:= n;
      while (not (j in thisset)) do j := j-1;
      last := j
    end else last := 0
end { last } ;

procedure swap(var a,b : integer);
{ exchanges a and b }
var t : integer;
begin t := a;a := b;b := t end;

function loop(new: integer; var i,j : arcarr;
var thisset : nodeset): boolean;
{  tests to see whether the addition
of the arc new
        to thisset creates a loop }
var l,lf,ls,la,lc : integer;
    t : array[1..maxnodes] of nodenos;
    lp : boolean;
    tset : nodeset;
begin
{variables used :
t    : vector of nodes found to be
connected to i[new]
tset : set of nodes in t (it is easy
to test for set membership)
la   : pointer (in t) to last node added
lc   : pointer (in t) to last node checked
ls   : pointer in list of all nodes to
the smallest member of thisset
lf   : pointer in list of all nodes to
the final member of thisset
lp   : true if j[new] in tset
}
    t[1] := i[new];
    tset := [i[new]];
    la := 1;
    lc := 1;
    lp := false;
    ls := least(maxnodes,thisset);
    lf := last(maxnodes,thisset);
    if (count(maxnodes,thisset))>0) then
  repeat
      for l := ls to lf do
        if ((not lp)and(l in thisset)) then
        begin
if (((t[lc]=i[l]))and not(j[l] in tset))
  then
            begin
              la := la+1;
              t[la] := j[l]];
              tset := tset + [j[l]];
```

```
            if (t[la]=j[new]) then lp := true
        end;
if ((t[lc]=j[l])and not(i[l] in tset))
  then
        begin
           la := la+1;
           t[la] := i[l];
           tset := tset + [i[l]];
           if (t[la]=j[new]) then lp := true
        end
     end;
     lc := lc+1
  until ((lc>la) or lp);
  loop := lp
end {loop};

begin
write(' Minimal spanning tree algorithm');
   writeln(':  Kruskals method');
   writeln;
   repeat
write('How many nodes, how many arcs?');
   readln(n,arcs);
   if (arcs<n-1) then
      writeln('There must be at least',
              n-1,' arcs. Try again')
   until (arcs >= n-1);
   write('Enter arcs with their');
writeln(' start node finish node length');
   for t8 := 1 to arcs do
      repeat
         write('Arc number ',t8:3,'  ');
         readln(i[t8],j[t8],d[t8]);
         if ((i[t8]>n)or(j[t8]>n)) then
writeln('Node number too high: try again');
         if ((i[t8]<0)or(j[t8]<0)) then
writeln('Node number too low: try again');
         if (i[t8]=j[t8]) then
writeln('start and finish must',
   ' be different: try again ')
      until ((i[t8]<>j[t8])and(i[t8]in[1..n])
and(j[t8]in[1..n]));
{              ******
               STEP 0
               ******
}
   s := [];
{              ************
               STEPS 1 AND 2
               ************
}
   for t8 := 1 to arcs do
   if (count(maxnodes,s)<(n-1)) then
      begin
         if (t8<arcs) then
         for t9 := t8+1 to arcs do
         if (d[t9]<d[t8]) then
         begin
           swap(i[t8],i[t9]);
           swap(j[t8],j[t9]);
           swap(d[t8],d[t9])
         end;
         if (not loop(t8,i,j,s))
then s := s+[t8]
      end;
```

```
{                  ******
                   STEP 3
                   ******
}
  if (count(maxnodes,s)<(n-1)) then
writeln('No tree exists')
    else
    begin
      writeln('Spanning tree');
      st := 0;
      for t8 := 1 to arcs do
      if (t8 in s) then
      begin
        st := st + d[t8];
        writeln(i[t8],j[t8],d[t8])
      end;
    writeln('This has length',st)
    end
end.
```

```
90 REM PROG2P2
100 REM KRUSKAL'S ALGORITHM
110 DIM I(30),J(30),D(30),L(30),M(30)
120 A9=30
130 N9=30
140 PRINT "MINIMAL SPANNING TREE
    ALGORITHM:   KRUSKAL'S METHOD"
150 PRINT
160 PRINT "HOW MANY NODES, HOW MANY ARCS?"
170 INPUT N,A
180 T5=0
190 IF A<N-1 GOSUB 860
200 IF A>A9 THEN 900
210 IF N>N9 THEN 940
220 IF T5>0 THEN 160
230 PRINT "ENTER ARCS WITH THEIR START
    NODE,FINISH NODE,LENGTH"
240 FOR T8=1 TO A
250 PRINT "ARC NUMBER ";T8;": ";
260 INPUT I(T8),J(T8),D(T8)
270 M(T8)=0
280 T5=0
290 IF I(T8)>N GOSUB 980
300 IF J(T8)>N GOSUB 980
310 IF I(T8)<=0 GOSUB 1010
320 IF J(T8)<=0 GOSUB 1010
330 IF I(T8)=J(T8) GOSUB 1040
340 IF D(T8)<0 GOSUB 1070
350 IF T5<>0 THEN 250
360 NEXT T8
370 K=0
380 A1=A-1
390 FOR T8=1 TO A1
400 T5=T8+1
410 FOR T9=T5 TO A
420 IF D(T9)>=D(T8) THEN 520
430 T7=I(T8)
440 I(T8)=I(T9)
450 I(T9)=T7
460 T7=J(T8)
470 J(T8)=J(T9)
```

```
480 J(T9)=T7
490 T7=D(T8)
500 D(T8)=D(T9)
510 D(T9)=T7
520 NEXT T9
530 FOR T9=1 TO A
540 L(T9)=0
550 NEXT T9
560 I2=I(T8)
570 J2=J(T8)
580 L(I2)=1
590 C=0
600 FOR T9=1 TO A
610 IF M(T9)<>1 THEN 690
620 I1=I(T9)
630 J1=J(T9)
640 IF L(I1)+L(J1)<>1 THEN 690
650 C=1
660 L(I1)=1


670 L(J1)=1
680 IF L(J2)=1 THEN 740
690 NEXT T9
700 IF C>0 THEN 590
710 M(T8)=1
720 K=K+1
730 IF K=N-1 THEN 770
740 NEXT T8
750 PRINT "THERE IS NO SPANNING TREE"
760 STOP
770 PRINT "THE SPANNING TREE IS"
780 S=0
790 FOR T8=1 TO A
800 IF M(T8)=0 THEN 830
810 PRINT I(T8),J(T8)
820 S=S+D(T8)
830 NEXT T8
840 PRINT "THIS HAS LENGTH";S
850 STOP
860   PRINT "THERE ARE NOT ENOUGH ARCS
      TO FORM A SPANNING TREE:"
870   PRINT "TRY AGAIN"
880   T5=1
890   RETURN
900   PRINT"THE PROGRAM HAS BEEN SET UP
      FOR AT MOST ";A9;" ARCS"
910   PRINT "IF YOU WANT MORE ARCS,
      PLEASE REDIMENSION ARRAYS "
920   PRINT "I,J,D,M AND CHANGE VARIABLE A9"
930   STOP
940   PRINT "THE PROGRAM HAS BEEN SET UP
      FOR AT MOST ";N9;" NODES"
950   PRINT "IF YOU WANT MORE NODES,
      PLEASE REDIMENSION ARRAY L"
960   PRINT "AND CHANGE VARIABLE N9"
970   STOP
980   PRINT "NODE NUMBER TOO HIGH:  TRY
      AGAIN"
990   T5=1
1000 RETURN
1010 PRINT "NODE NUMBER TOO LOW:  TRY AGAIN"
1020 T5=1
1030 RETURN
```

```
1040 PRINT "START AND FINISH IDENTICAL
     :TRY AGAIN"
1050 T5=1
1060 RETURN
1070 PRINT "ARC LENGTH SHOULD BE
     POSITIVE:   TRY AGAIN"
1080 T5=1
1090 RETURN
```

Implementing Prim's algorithm

The key part of Prim's algorithm is step 1, where one of the criteria for selecting the next arc for inclusion is that the arc should have one node in the partial tree and the other not in it. This requires that the computer program should have some means of identifying the nodes which are in the tree. In PASCAL this can be done with a vector of Boolean variables, which will be false for nodes not yet connected to the partial tree, and true for those nodes which are so connected. Then, for arcs which are suitable candidates for inclusion, it is necessary to check that one node has a true label, and the other has a false label. In the program this is done with a Boolean function, 'diff', which is true when exactly one of the nodes has a true label. In BASIC, the same principle is applied, but the labels are numerical, either -1 or $+1$. Then the sum of the labels on the ends of an arc can be tested; eligible arcs will have this sum equal to 0. In each program, the inclusion of an arc means that both its ends become labelled; this saves the need to pick out which one was already labelled.

In both the BASIC and the PASCAL programs, the first arc to be included is not chosen arbitrarily, as the formal algorithm suggests. Instead, while the distance matrix is being input, the program stores the shortest arc, and this is used as the first arc to be added to the tree. The effect of this minor deviation is to speed up the search of step 1 in most cases.

Both programs need to detect networks without spanning trees. The formal algorithm assumes that the network is connected, and when using the algorithm by hand, it is a trivial matter to detect the occasions when this assumption is not justified. However, it is less straightforward for the computer, although there are algorithms which can be used to check this, two of which are given later. In the programs here, a check is made to see whether the pass through step 1 finds any eligible arcs for the partial tree; if no such arc is found, then the variable 'minarc' (PASCAL) or 'M1' (BASIC) is not changed by the step. A warning message is printed if this happens. The completeness of the tree is checked by counting the number of nodes not included in the tree.

Finally, there has to be a way of storing the tree in a convenient form, to allow it to be recovered for display. This could be done with a Boolean matrix, matching the distance matrix (or an equivalent matrix in BASIC), but it is more convenient to use the distance matrix itself to store this information. Those arcs which are included in the tree have their distance replaced by -1; should the distance matrix be required for later use, then the entry in the (i, j)th position could be set to its negative complement, until the tree was output, and then returned to its true value.

```
program prog2p3(input,output);
  const inf = 9999;
        maxnodes = 20;
  var   nodes,i,j,t6,t7,nnintree,i1,j1,minarc,
##    st : integer;
        d : array[1..maxnodes,1..maxnodes] of
##    integer;
        l : array[1..maxnodes] of boolean;

  function diff(b1,b2 : boolean) : boolean;
  begin diff := ((b1 and not b2) or (b2 and
##    not b1)) end ;
  { diff is true if b1 is different from b2,
##    false if they are the same }

  begin
  {     Minimal Spanning Tree Algorithm, using
##    Prim's method
  (the 'greedy algorithm')
  variables used
  nodes       : number of nodes
  i,j,t6,t7 : work space
  nnintree      : number of nodes not in the tree
  i1,j1 : ends of the next arc to be added to
##    the tree
  minarc: length of the next arc to be added
  st      : total length of the tree, as so far
##    created
  d       : matrix of distances
  l       : boolean vector for nodes true when
##    a node is in the tree
  }
    writeln('      The Minimum Spanning Tree
##    Algorithm');
    writeln('      Prims method'); writeln;
  write('How many nodes in the network?  ');
    readln(nodes);
    if (nodes>maxnodes) then
    begin
        writeln('The program has been set up
##    for networks with at most',maxnodes);
        writeln('nodes.    If you want to use a
##    larger network, please reset the');
  writeln('constant maxnodes in the program');
        writeln
    end
    else
    begin
        for i := 1 to nodes do
        begin
            for j := 1 to nodes do  d[i,j] := inf;
##    l[i] := false
        end ;
        writeln(' Enter arcs in the form: start
##    node finish node length');
        writeln(' Enter 0 0 0 to finish');
        t6 := 0;
        minarc := inf;
```

```
       repeat
         t6 := t6+1;
         repeat
           write('Arc number ',t6:3,' ');
           readln(i,j,t7);
           if (t7 < 0) then writeln('Length
##    must be positive.    Try again');
      if ((i>nodes)or(j>nodes)) then
##    writeln('Node number too high.    Try again');
           if ((i<0)or(j<0)) then
##    writeln('Node number too low.    Try again')
           until ((i in [0..nodes])and(j in
##    [0..nodes])and(t7>=0));
           if (i<>0) then
           begin
             d[i,j] := t7;
             if (t7<minarc) then
             begin
               minarc := t7;
               i1 := i;
               j1 := j;
             end
           end
         until (i=0);

         writeln('The matrix has been input as
##    below:');
         for i := 1 to nodes do
         begin
           for j := 1 to nodes do  write(d[i,
##    j]);  writeln
         end ;

{                ******
                 STEP 0
                 ******
}
         l[i1] := true;
         l[j1] := true;
         d[i1,j1] := -1;
         st := minarc;
         nnintree := nodes-2;

         repeat
           minarc := inf;
           for i := 1 to nodes do
           for j := 1 to nodes do
{                ******
                 STEP 1
                 ******
}
           if ((diff(l[i],l[j])) and (d[i,
##    j]<minarc)) then
             begin
               minarc := d[i,j];
               i1 := i;
               j1 := j;
             end;
           l[i1] := true;
           l[j1] := true;
           st := st+minarc;
           d[i1,j1] := -d[i1,j1];
           nnintree := nnintree-1
         until ((minarc = inf)or(nnintree=0));
```

```
(                     ******
                      STEP 2
                      ******
)
        if (minarc = inf) then writeln('There
##    is no spanning tree')
        else
        begin
writeln('The minimum spanning tree is ');
        for i := 1 to nodes do
        for j := 1 to nodes do
if (d[i,j]<0) then writeln(i:4,'  ',j:4);

        writeln('This has length ',st)
      end
    end
 end.
```

```
90 REM PROG2P4
100 REM PRIM'S METHOD
110 DIM D(20,20),L(20)
120 REM VARIABLES USED
130 REM D: DISTANCE MATRIX;  D(I,J)=ARC
    LENGTH I TO J
140 REM   D(I,J) IS SET TO -1 WHEN (I,
    J) IS IN THE SPANNING TREE
150 REM L: LABEL ON NODES:  L(I)=-1
    WHEN I IS NOT IN TREE
160 REM   L(I)=+1 WHEN I IS IN TREE
170 REM I,J: LOOP COUNTERS
180 REM N: NUMBER OF NODES
190 REM S: LENGTH OF TREE CREATED SO FAR
200 REM M1: LENGTH OF SHORTEST ARC
    WHICH MAY BE ADDED TO TREE
210 REM (I1,J1):  THE ARC WHOSE LENGTH IS M1
220 REM I9:   MAXIMUM NUMBER OF NODES
230 REM T9:   INFINITY
240 REM T8:   NUMBER OF ARCS NEEDED TO
    COMPLETE THE TREE
250 REM T7:  USED FOR STORING ARC
    LENGTHS ON INPUT
260 REM T6:  USED FOR COUNTING ARCS
270 REM T5:  USED FOR CHECKING INPUT
280 I9=20
290 T9=9999
300 PRINT "THE MINIMUM SPANNING TREE
    ALGORITHM,  PRIM'S METHOD"
310 PRINT
320 PRINT "HOW MANY NODES IN THE NETWORK? ";
330 INPUT N
340 IF N <= I9 THEN 390
350 PRINT "THE PROGRAM HAS BEEN SET UP
    FOR NETWORKS WITH AT"
360 PRINT "MOST ";I9;" NODES,  IF YOU
    WISH TO USE A LARGER"
370 PRINT "NETWORK, PLEASE INCREASE THE
    DIMENSIONS AND I9"
380 STOP
390 FOR I=1 TO N
400 FOR J=1 TO N
410 D(I,J)=T9
420 NEXT J
430 L(I)=-1
```

```
440 NEXT I
450 PRINT "ENTER THE ARCS IN THE FORM
    START NODE,FINISH NODE,LENGTH"
460 PRINT "ENTER 0,0,0 TO FINISH"
470 T6 = 0
480 M1=T9
490 T6=T6+1
500 PRINT "ARC NUMBER";T6;": ";
510 INPUT I,J,T7
520 T5=0
530 IF T7<0 GOSUB 1170
540 IF I>N GOSUB 1200
550 IF J>N GOSUB 1200
560 IF I<0 GOSUB 1230
570 IF J<0 GOSUB 1230


580 IF T5>0 THEN 500
590 IF I=0 THEN 660
600 D(I,J)=T7
610 IF T7>=M1 THEN 490
620 M1 = T7
630 I1=I
640 J1=J
650 GOTO 490
660 PRINT "THE MATRIX HAS BEEN INPUT AS
    BELOW"
670 FOR I = 1 TO N
680 FOR J= 1 TO N
690 PRINT D(I,J);
700 NEXT J
710 PRINT
720 NEXT I
730 IF T6<N THEN 1260
740 L(I1)=1
750 L(J1)=1
760 D(I1,J1)=-1
770 S=M1
780 T8=N-2
790 REM          ******
800 REM          STEP 0
810 REM          ******
820 M1=T9
830 FOR I=1 TO N
840 FOR J=1 TO N
850 REM          ******
860 REM          STEP 1
870 REM          ******
880 IF L(I)+L(J)<>0 THEN 930
890 IF D(I,J)>=M1 THEN 930
900 M1=D(I,J)
910 I1=I
920 J1=J
930 NEXT J
940 NEXT I
950 L(I1)=1
960 L(J1)=1
970 S=S+M1
980 D(I1,J1)=-1
990 T8=T8-1
1000 IF T8=0 THEN 1060
1010 IF M1=T9 THEN 1150
1020 GOTO 820
```

```
1030 REM                ******
1040 REM                STEP 2
1050 REM                ******
1060 PRINT "THE MINIMUM SPANNING TREE IS"
1070 FOR I = 1 TON
1080 FOR J=1 TO N
1090 IF D(I,J)>=0 THEN 1110
1100 PRINT I,J
1110 NEXT J
1120 NEXT I
1130 PRINT "THIS HAS LENGTH ";S
1140 STOP
1150 PRINT "THERE IS NO SPANNING TREE"
1160 STOP


1170 PRINT "ARC LENGTH MUST BE
     POSITIVE:  TRY AGAIN"
1180 T5=1
1190 RETURN
1200 PRINT "NODE NUMBER TOO HIGH:   TRY
     AGAIN"
1210 T5=1
1220 RETURN
1230 PRINT "NODE NUMBER TOO LOW:  TRY AGAIN"
1240 T5=1
1250 RETURN
1260 PRINT "THERE ARE NOT ENOUGH ARCS
     TO FORM A SPANNING TREE"
1270 STOP
```

2.3 CONNECTEDNESS: ALGORITHM 1

The idea of connectedness in a graph has been introduced earlier, and passing reference made to it in the description of the computer programs in this chapter. Although it is very easy to determine from a diagram whether or not a given graph is connected, it is more difficult when the components of the network are given in numerical form for computational use. However, there are several algorithms which can be used for determining whether or not a graph is connected.

The simplest, though not the most efficient, uses the adjacency matrix, X, of the graph. This is the $n * n$ matrix defined in the first chapter, with entries which are zero or one. If this matrix is squared, then its entries correspond to the number of different paths between two nodes which use 2 arcs. If it is cubed, then the entry in the (i,j)th position is the number of different paths between the nodes i and j which use 3 arcs; raised to the fourth power, it yeilds a matrix of the number of ways between two nodes using 4 arcs. If a graph is connected, then it is possible to find a path between every pair of nodes which uses at most n-1 arcs. So at least one of the matrices $X, X^2 \ldots X^{n-1}$ will have a non-zero entry in the (i,j)th position if a path between i and j exists. In order to test whether a graph is connected, the sum $Y = X + X^2 + X^3 + \ldots + X^{n-1}$ is found. This will have zero entries for any nodes which are not connected; if this matrix is wholly non-zero, then the graph is connected.

Formally, this may be stated as an algorithm:

step 0   Set $Y = X$.
          Set $k = 1$.
step 1   If Y has a row which is free from zeroes, then do step 3.
step 2   Set $k = k + 1$.
          If $k = n$, do step 4.
          Set $Y = Y + X^k$; do step 1.
step 3   Stop – the graph is connected.
step 4   Stop – the graph is not connected.

2.3.1 Worked Example

Consider the following graph and adjacency matrix (Fig. 2.7) (it is obvious that this is not connected).

$$\begin{bmatrix} 0 & 1 & 0 & 0 & 0 \\ 1 & 0 & 0 & 0 & 1 \\ 0 & 0 & 0 & 1 & 0 \\ 0 & 0 & 1 & 0 & 0 \\ 0 & 1 & 0 & 0 & 0 \end{bmatrix}$$

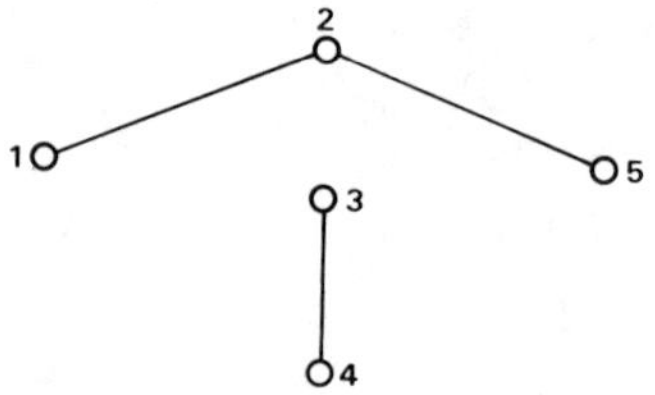

Fig. 2.7.

step 0 $Y = \begin{bmatrix} 0 & 1 & 0 & 0 & 0 \\ 1 & 0 & 0 & 0 & 1 \\ 0 & 0 & 0 & 1 & 0 \\ 0 & 0 & 1 & 0 & 0 \\ 0 & 1 & 0 & 0 & 0 \end{bmatrix}$

          $k = 1$

step 1 Y has no rows free of zeroes.

step 2 $k = 2 \neq 5$; $X^2 = \begin{bmatrix} 1 & 0 & 0 & 0 & 1 \\ 0 & 2 & 0 & 0 & 0 \\ 0 & 0 & 1 & 0 & 0 \\ 0 & 0 & 0 & 1 & 0 \\ 1 & 0 & 0 & 0 & 1 \end{bmatrix}$

          $Y = \begin{bmatrix} 1 & 1 & 0 & 0 & 1 \\ 1 & 2 & 0 & 0 & 1 \\ 0 & 0 & 1 & 1 & 0 \\ 0 & 0 & 1 & 1 & 0 \\ 1 & 1 & 0 & 0 & 1 \end{bmatrix}$

step 1 Y has no rows free of zeroes.

step 2 $k = 3 \neq 5$; $X^3 = \begin{bmatrix} 0 & 2 & 0 & 0 & 0 \\ 2 & 0 & 0 & 0 & 2 \\ 0 & 0 & 0 & 1 & 0 \\ 0 & 0 & 1 & 0 & 0 \\ 0 & 2 & 0 & 0 & 0 \end{bmatrix}$

$Y = \begin{bmatrix} 1 & 3 & 0 & 0 & 3 \\ 3 & 2 & 0 & 0 & 3 \\ 0 & 0 & 1 & 2 & 0 \\ 0 & 0 & 2 & 1 & 0 \\ 1 & 3 & 0 & 0 & 1 \end{bmatrix}$

step 1 Y has no rows free from zeroes.

step 2 $k = 4 \neq 5$; $X^4 = \begin{bmatrix} 2 & 0 & 0 & 0 & 2 \\ 0 & 4 & 0 & 0 & 0 \\ 0 & 0 & 1 & 0 & 0 \\ 0 & 0 & 0 & 1 & 0 \\ 2 & 0 & 0 & 0 & 2 \end{bmatrix}$

$Y = \begin{bmatrix} 3 & 3 & 0 & 0 & 3 \\ 3 & 6 & 0 & 0 & 3 \\ 0 & 0 & 2 & 2 & 0 \\ 0 & 0 & 2 & 2 & 0 \\ 3 & 3 & 0 & 0 & 3 \end{bmatrix}$

step 1 Y has no rows free from zeroes.
step 2 $k = 5$; do step 4.
step 4   the graph is not connected.

2.3.2 Note

If one wanted to know the different components of the graph, then the final value of Y provides a means of recognising these. Node 1 is connected by a chain to all the nodes for which $Y_{1j} \neq 0$; in the case above, these nodes are 1, 2 and 5. The next component of the graph can be found by selecting one of the nodes which has not been listed and finding which nodes are connected to it; so node 3 might be selected, and it will be seen that node 4 is connected to it.

2.3.3 Implementation

This algorithm presents few difficulties for implementation; it is useful to store a matrix Z equal to X^k, rather than calculate this every time. In some versions of BASIC there is a procedure for multiplying two matrices available, but in the program presented here, this is done with a small subroutine. The test on the rows of Y is carried out by scanning until a zero is found in each row. If the whole row is scanned and no zero is found, then the graph will have been found to be connected; no further checks are needed.

```
program prog2p5(input,output);
  const maxnodes = 20;
  type matrix = array[1..maxnodes,
##    1..maxnodes] of integer;
  var x,y,z : matrix;
      c,i,j,k,nodes : integer;
      connect : boolean;

  procedure matmult(var a,b : matrix ; n :
##    integer);
  {sets a to be the product of a and b }
  var i,j,k : integer;
      c : matrix;
  begin
    for i := 1 to n do
    for k := 1 to n do
    begin
      c[i,k] := 0;
      for j := 1 to n do c[i,k] := c[i,k] +
##    a[i,j]*b[j,k]
    end;
    for i := 1 to n do
    for j := 1 to n do
      a[i,j] := c[i,j]
  end  {matmult};

  procedure matadd(var a,b : matrix; n :
##    integer);
  {sets a to be the sum of a and b }
  var i,j : integer;
  begin
    for i := 1 to n do
    for j := 1 to n do
      a[i,j] := a[i,j] + b[i,j]
  end {matadd};

  begin
  {connectedness algorithm, using powers of
##    the adjacency matrix

  variables used:
  x     : the adjacency matrix
  z     : the latest power of the adjacency
##    matrix to be calculatd
  y     : the sum of the powers of x
  c     : counter for the number of arcs (on
##    exit from the loop equals 1 more
          than the number of arcs)
  nodes : number of nodes in the graph
  i,j,k : loop variables
  connect : variable which becomes true when
##    the graph is found to be connected
  }

  repeat
    write('How many nodes?  ');
    readln(nodes);
    if (nodes <= 0) then writeln('There must
##    be at least 1 node;  try again')
  until (nodes > 0);
```

```
    if (nodes>maxnodes) then
    begin
      writeln('The program has been set up for
##      networks with at most',maxnodes:4);
      writeln('nodes.   If you wish to use'
##      larger networks, please change the');
      writeln('parameter maxnodes')
    end
    else
    begin
      for i := 1 to nodes do
      for j := 1 to nodes do
        x[i,j] := 0;
      writeln('Enter arcs now, giving their two
##      end nodes.   Enter 0 0 to stop');
      c := 0;
      repeat
        c := c+1;
        repeat
          write('Arc number',c:4,': ');
          readln(i,j);
          if ((i>nodes)or(j>nodes))then
##        writeln('Node number too high; try again');
          if ((i<0)or(j<0)) then writeln('Node
##        number too low; try again')
        until ((i in [0..nodes]) and (j in
##        [0..nodes]));
        if (i>0) then
        begin
          x[i,j] := 1;
          x[j,i] := 1
        end
      until (i=0);
      writeln;
      if (c<nodes) then
      begin
        writeln('This graph cannot be connected
##      since there are not enough');
writeln('arcs to form a spanning tree.')
      end;
    {           ******
                STEP 0
                ******
    }
      for i := 1 to nodes do
      for j := 1 to nodes do
      begin
        z[i,j] := x[i,j];
        v[i,j] := x[i,j]
      end;
      connect := false;
      k := 1;
    {           ******
                STEP 1
                ******
    }
      repeat
        i := 0;
        repeat
          i := i+1;
          j := 0;
          repeat
            j := j+1;
          until ((j>=nodes)or(v[i,j]=0));
          if ((j=nodes)and(v[i,nodes]<>0)) then
```

```
##      connect := true
        until ((i)=nodes)or connect);
        if not connect then
        begin
{               ******
                STEP 2
                ******
}
        k := k+1;
        if (k<nodes) then
        begin
           matmult(z,x,nodes);
           matadd(y,z,nodes)
        end
     end
   until ((k=nodes) or connect);
{               *************
                STEPS 3 AND 4
                *************
}
  if connect then
    writeln('This graph is connected')
  else
  begin
    writeln('This graph is not connected');
    writeln;
    writeln('The final matrix Y is:');
    writeln;
    for i := 1 to nodes do
    begin
       for j := 1 to nodes do
         write(y[i,j]:6);
       writeln
    end
  end
 end
 end.
```

```
90 REM PROG2P6
100 REM CONNECTEDNESS USING POWERS OF
    ADJACENCY MATRIX
110 REM
120 REM VARIABLES USED
130 REM
140 REM X    :ADJACENCY MATRIX
150 REM    :X RAISED TO THE POWER K
160 REM Y    : SUM OF POWERS OF X
170 REM A    : WORKSPACE
180 REM N9   : MAXIMUM NUMBER OF NODES
190 REM N    : NUMBER OF NODES
200 REM I,J  : LOOP COUNTERS
210 REM K    :POWER OF X
220 REM C    : NUMBER OF ARCS (DEFINED
    AS ONE LESS THAN C)
230 REM
240 DIM X(20,20),Y(20,20),Z(20,20),A(20,20)
250 N9=20
260 PRINT "CONNECTEDNESS USING POWERS
    OF ADJACENCY MATRIX"
270 PRINT
280 PRINT "HOW MANY NODES ? :";
290 INPUT N
300 IF N<=0 THEN 360
```

```
310 IF N<=N9 THEN 380
320 PRINT "THE PROGRAM HAS BEEN SET UP
    FOR NETWORKS WITH AT MOST";N9
330 PRINT "NODES.   IF YOU WISH TO USE
    LARGER NETWORKS, PLEASE CHANGE"
340 PRINT "THE DIMENSIONS OF ARRAYS AND
    THE PARAMETER N9"
350 STOP
360 PRINT "THERE MUST BE AT LEAST ONE
    NODE:  TRY AGAIN"
370 GOTO 280
380 FOR I=1 TO N
390 FOR J=1 TO N
400 X(I,J)=0
410 NEXT J
420 NEXT I
430 PRINT "ENTER ARCS NOW, GIVING THEIR
    TWO END NODES"
440 PRINT "ENTER 0,0 TO STOP"
450 C=0
460 C=C+1
470 PRINT "ARC NUMBER ";C;":   ";
480 INPUT I,J
490 T5=0
500 IF I>N GOSUB 1250
510 IF J>N GOSUB 1250
520 IF I<0 GOSUB 1280
530 IF J<0 GOSUB 1280
540 IF T5>0 THEN 470
550 IF I=0 THEN 590
560 X(I,J)=1
570 X(J,I)=1
580 GOTO 460
590 PRINT
600 IF C>=N THEN 660


610 PRINT "THIS GRAPH CANNOT BE
    CONNECTED SINCE THERE ARE NOT ENOUGH"
620 PRINT "ARCS TO FORM A SPANNING TREE"
630 REM               ******
640 REM               STEP 0
650 REM               ******
660 FOR I=1 TO N
670 FOR J=1 TO N
680 Y(I,J)=X(I,J)
690 Z(I,J)=X(I,J)
700 NEXT J
710 NEXT I
720 K=1
730 REM               ******
740 REM               STEP 1
750 REM               ******
760 I=0
770 I=I+1
780 J=0
790 J=J+1
800 IF J>N THEN 910
810 IF Y(I,J)=0 THEN 830
820 GOTO 790
830 IF I<N THEN 770
840 REM               ******
850 REM               STEP 2
860 REM               ******
```

```
 870 K=K+1
 880 IF K=N THEN 960
 890 GOSUB 1080
 900 GOTO 760
 910 REM              ******
 920 REM              STEP 3
 930 REM              ******
 940 PRINT "THIS GRAPH IS CONNECTED"
 950 STOP
 960 REM              ******
 970 REM              STEP 4
 980 REM              ******
 990 PRINT "THIS GRAPH IS NOT CONNECTED"
1000 PRINT "THE FINAL MATRIX Y IS"
1010 FOR I=1 TO N
1020 FOR J=1 TO N
1030 PRINT Y(I,J);
1040 NEXT J
1050 PRINT
1060 NEXT I
1070 STOP
1080 REM SUBROUTINE TO MULTIPLY Z BY X,
     AND STORE THE RESULT IN Z
1090 REM THEN TO ADD Z TO Y
1100 FOR P=1 TO N
1110 FOR R=1 TO N
1120 A(P,R)=0
1130 FOR Q=1 TO N
1140 A(P,R)=A(P,R) + X(P,Q)*Z(Q,R)
1150 NEXT Q
1160 NEXT R
1170 NEXT P
1180 FOR P=1 TO N


1190 FOR Q=1 TO N
1200 Z(P,Q)=A(P,Q)
1210 Y(P,Q)=Y(P,Q)+Z(P,Q)
1220 NEXT Q
1230 NEXT P
1240 RETURN
1250 PRINT "NODE NUMBER TOO HIGH:   TRY
     AGAIN"
1260 T5=1
1270 RETURN
1280 PRINT "NODE NUMBER TOO LOW:   TRY AGAIN"
1290 T5=1
1300 RETURN
```

2.4 CONNECTEDNESS: ALGORITHM II

The main drawback to the algorithm which has just been described is that it requires a great deal of matrix multiplication, which, for a large graph, would be very time-consuming. A much more efficient algorithm is available, which creates a 'connectedness matrix' by addition of rows and columns of the adjacency matrix. The connectedness matrix is similar to the adjacency matrix, with zero/one entries; in the ith row, there will be an entry of +1 for each node to which i is connected, irrespective of the length of the path needed to reach it. This matrix is built up from the adjacency matrix using the principle: if i is connected to j, and j to k, then i is connected to k. This can be expressed in terms of logical addition of the jth row to the ith row, and the jth column to the ith column, followed by deletion of the jth row and column. In this addition, the rules are: $0 + 0 = 0$, $1 + 0 = 0 + 1 = 1 + 1 = 1$. To test for connectedness the rules are applied to node 1, until either all the other rows and columns have been deleted, or no other ones can be deleted. If the former, then the graph is connected; if not, then repeating the rules for the remaining nodes will eventually reveal how many components there are in the graph.

Formally, this may be stated as an algorithm:

step 0   Initialise X as the adjacency matrix.
          Set $i = 1$ $c = 0$.
step 1   Find $X_{ij} \neq 0$ $(j > i)$.
          Logically add row j to row i, column j to column i and delete row j and column j from X.
          If no such j exists then do step 2 otherwise repeat step 1.
step 2   Set $c = c + 1$.
          Find $k > i$ such that row k has not been deleted; if no such k exists then stop, with a graph of c components; otherwise set $i = k$ and do step 1.

2.4.1 Worked Example

Using the same graph as before, namely,

$$X = \begin{bmatrix} 0 & 1 & 0 & 0 & 0 \\ 1 & 0 & 0 & 0 & 1 \\ 0 & 0 & 0 & 1 & 0 \\ 0 & 0 & 1 & 0 & 0 \\ 0 & 1 & 0 & 0 & 0 \end{bmatrix}$$

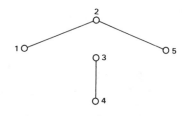

Fig. 2.8.

step 0 X initialised as above.
          $i = 1$, $c = 0$.

step 1 $X_{12} = 1 \neq 0$
          adding the 2nd row to the 1st and the 2nd column to the 1st gives:

$$X = \begin{bmatrix} 1 & 1 & 0 & 0 & 1 \\ 1 & 0 & 0 & 0 & 1 \\ 0 & 0 & 0 & 1 & 0 \\ 0 & 0 & 1 & 0 & 0 \\ 1 & 1 & 0 & 0 & 0 \end{bmatrix}$$

which becomes:

$$X = \begin{bmatrix} 1 & - & 0 & 0 & 1 \\ - & - & - & - & - \\ 0 & - & 0 & 1 & 0 \\ 0 & - & 1 & 0 & 0 \\ 1 & - & 0 & 0 & 0 \end{bmatrix}$$

when the 2nd row and column have been deleted.

step 1 $X_{15} = 1 \neq 0$,

adding the 5th row to the 1st and 5th column to the 1st gives:

$$X = \begin{bmatrix} 1 & - & 0 & 0 & - \\ - & - & - & - & - \\ 0 & - & 0 & 1 & - \\ 0 & - & 1 & 0 & - \\ - & - & - & - & - \end{bmatrix}$$

after deletion of the 5th row and column.

step 1   There is no other $X_{1j} \neq 0$.

step 2 $c = 1$

$k = 3; i = 3$

step 1 $X_{34} = 1 \neq 0$,

adding the 4th row and column to the 3rd gives:

$$X = \begin{bmatrix} 1 & - & 0 & - & - \\ - & - & - & - & - \\ 0 & - & 1 & - & - \\ - & - & - & - & - \\ - & - & - & - & - \end{bmatrix}$$

after deletion.

step 1   There is no $X_{3j} \neq 0$ remaining.

step 2 $c = 2$,

no k exists, so the algorithm stops with 2 components.

2.4.2 Implementation

In practice, it may be desirable not only to record the number of components of the graph, but also the nodes which go to make these up. This information could be output from the computer program as step 1 was performed, but it is more convenient to store it within the program. A simple way of doing this is to record, in step 1, the i which corresponds to each j in a vector of integers. This also would enable the rows and columns of the matrix to be identified; if a given row had a label then it would have been deleted from the matrix; if not then it will remain a suitable candidate for steps 1 and 2. Apart from this consideration, the programming is a straightforward translation of the algorithm, with the adjacency matrix being held as a square matrix, and gradually converted into the connectedness matrix.

```
program prog2p7(input,output);
  const maxnodes = 20;
  type matrix = array[1..maxnodes,
##      1..maxnodes] of integer;
  var x : matrix;
       c,comp,i,j,k,nodes : integer;
       nochange : boolean;
       l : array[1..maxnodes] of integer;

  function max(a,b : integer) : integer ;
  { calculates the maximum of two integers,
##      which is equivalent to
     logical addition of the integers}
  begin
     if (a>b) then max := a else max := b
  end { max } ;

  begin

  {connectedness algorithm, using logical
##      addition of rows and columns

  variables used:
     x       : the adjacency matrix, converted into
##       the connectedness matrix
     c       : counter for the number of arcs (on
##      exit from the loop equals 1 more
                  than the number of arcs)
     comp : counter for the number of components
##      completed
  nodes : number of nodes in the graph
  i,j,K : loop variables, used in scanning
##       the matrix
  }
  repeat
     write('How many nodes?   ');
     readln(nodes);
     if (nodes <= 0) then
        writeln('There must be at least 1 node:
##        try again')
  until (nodes > 0);

  if (nodes>maxnodes) then
  begin
     writeln('The program has been set up for
##      networks with at most',maxnodes:4);
     writeln('nodes.   If you wish to use
##       larger networks, please change the');
     writeln('parameter maxnodes')
  end
  else
  begin
     for i := 1 to nodes do
     for j := 1 to nodes do
        x[i,j] := 0;
     writeln('Enter arcs now, giving their two
##       end nodes,  Enter 0 0 to stop');
     c := 0;
```

```
      repeat
        c := c+1;
        repeat
          write('Arc number',c:4,': ');
          readln(i,j);
          if ((i>nodes)or(j>nodes))then
writeln('Node number too high: try again');
          if ((i<0)or(j<0)) then
writeln('Node number too low:  try again')
        until ((i in [0..nodes]) and (j in
##      [0..nodes]));
        if (i>0) then
        begin
          x[i,j] := 1;
          x[j,i] := 1
        end
      until (i=0);
      writeln;
      if (c<nodes) then
      begin
        writeln('This graph cannot be connected
##      since there are not enough');
writeln('arcs to form a spanning tree.')
      end;
  {             ******
                STEP 0
                ******
  }
      i := 1;
      comp := 0;
      for j := 2 to nodes do
        l[j] := 0;
      repeat
  {             ******
                STEP 1
                ******
  }
        l[i] := i;
  {   i is the lowest numbered node in the
##      current component }
        repeat
          j := i+1;
          nochange := true;
          repeat
  {   test to see whether j is in the current
##      component }
            if ((x[i,j]=1)and(l[j]=0)) then
            begin
              l[j] := i;
              for k := 1 to nodes do
              if (l[k]=0) then
              begin
                x[i,k] := max(x[i,k],x[j,k]);
                x[k,i] := x[i,k];
                nochange := false
              end
            end;
            j := j+1
          until (j>nodes)
        until nochange;
  {             ******
                STEP 2
                ******
  }
        comp := comp+1;
        k := i;
```

```
      repeat
        K := K+1
      until ((K>=nodes)or(l[k]=0));
      if (l[k]=0) then i := k
    until (i<>k);
    if (comp=1) then writeln('This graph is
##     connected')
    else
    begin
      writeln('This graph is not connected');
      writeln('It contains ',comp:4,'
##     components, which are:');
      for k := 1 to nodes do
      if (l[k]=k) then
      begin
        for j := k to nodes do
          if (l[j]=k) then write(j:4);
        writeln
      end
    end
  end
end.
```

```
90 REM PROG2P8
100 REM CONNECTEDNESS BY LOGICAL
    ADDITION OF ROWS AND COLUMNS
110 REM OF THE ADJACENCY MATRIX TO GIVE
    CONNECTEDNESS MATRIX
120 REM
130 REM VARIABLES USED
140 REM
150 REM X   : ADJACENCY MATRIX, CHANGED
    TO CONNECTEDNESS MATRIX
160 REM L   : VECTOR USED TO IDENTIFY
    THE NODES WHICH HAVE BEEN
170 REM         LINKED INTO A COMPONENT
    OF THE GRAPH
180 REM N9  : MAXIMUM NUMBER OF NODES
190 REM N   : NUMBER OF NODES
200 REM C   : NUMBER OF ARCS
210 REM T5  : TEST FOR LEGITIMACY OF DATA
220 REM C1  : COUNT OF COMPONENTS
230 REM I,J,K : LOOP COUNTERS
240 REM N1  : TEST FOR CHANGE IN X
    FOLLOWING A PASS THROUGH NODES
250 DIM X(20,20),L(20)
260 N9=20
270 PRINT "HOW MANY NODES ? : ";
280 INPUT N
290 IF N>N9 THEN 330
300 IF N>0 THEN 370
310 PRINT "THERE MUST BE AT LEAST ONE
    NODE: TRY AGAIN"
320 GOTO 270
330 PRINT "THE PROGRAM HAS BEEN SET UP
    FOR NETWORKS WITH AT MOST";N9
340 PRINT "NODES.  IF YOU WISH TO USE
    LARGER NETWORKS, THEN PLEASE"
350 PRINT "CHANGE THE PARAMETER N9 AND
    THE DIMENSIONS OF X AND L"
360 STOP
370 FOR I=1 TO N
```

```
380 FOR J=1 TO N
390 X(I,J)=0
400 NEXT J
410 NEXT I
420 PRINT "ENTER ARCS NOW, GIVING THEIR
    TWO END NODES"
430 PRINT "ENTER 0,0 TO STOP"
440 C=0
450 C=C+1
460 PRINT "ARC NUMBER ";C;":    ";
470 INPUT I,J
480 T5=0
490 IF I>N GOSUB 1120
500 IF J>N GOSUB 1120
510 IF I<0 GOSUB 1150
520 IF J<0 GOSUB 1150
530 IF T5>0 THEN 460
540 IF I=0 THEN 580
550 X(I,J)=1
560 X(J,I)=1
570 GOTO 450
580 IF C>=N THEN 610
590 PRINT "THIS GRAPH CANNOT BE
    CONNECTED SINCE THERE ARE NOT ENOUGH"
600 PRINT "ARCS TO FORM A SPANNING TREE"
610 REM              ******
620 REM              STEP 0
630 REM              ******
640 I=1
650 C1=0
660 FOR J=2 TO N
670 L(J)=0
680 NEXT J
690 REM              ******
700 REM              STEP 1
710 REM              ******
720 L(I)=I
730 J=I+1
740 N1=1
750 IF X(I,J)=0 THEN 840
760 IF L(J)<>0 THEN 840
770 L(J)=I
780 FOR K=1 TO N
790 IF L(K)<>0 THEN 830
800 X(I,K)=MAX(X(I,K),X(I,K),X(I,K),X(J,K))
810 X(K,I)=X(I,K)
820 N1=0
830 NEXT K
840 J=J+1
850 IF J<=N THEN 750
860 IF N1=0 THEN 730
870 REM              ******
880 REM              STEP 2
890 REM              ******
900 C1=C1+1
910 K=I
920 K=K+1
930 IF L(K)=0 THEN 960
940 IF K>=N THEN 980
950 GOTO 920
960 I=K
970 GOTO 720
980 IF C1>1 THEN 1010
990 PRINT "THIS GRAPH IS CONNECTED"
1000 STOP
1010 PRINT "THIS GRAPH IS NOT CONNECTED"
```

```
1020 PRINT "IT CONTAINS ";C1;"
     COMPONENTS WHICH ARE"
1030 FOR K=1 TO N
1040 IF L(K)<>K THEN 1100
1050 FOR J=K TO N
1060 IF L(J)<>K THEN 1080
1070 PRINT J;
1080 NEXT J
1090 PRINT
1100 NEXT K
1110 STOP
1120 PRINT "NODE NUMBER TOO HIGH:  TRY
     AGAIN"
1130 T5=1
1140 RETURN


1150 PRINT "NODE NUMBER TOO LOW:  TRY AGAIN"
1160 T5=1
1170 RETURN
```

EXERCISES

(2.1)   For the network shown, find the minimal spanning trees, using Kruskal's
         algorithm, and ranking the arcs of equal length in either possible order.

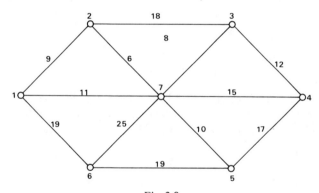

Fig. 2.9.

(2.2)   For the network of exercise (2.1), use Prim's algorithm to find the
         minimal spanning trees, using each node in turn as the first node of step 0.

(2.3)   In the network (N,A,D), the length of each arc $d(i,j)$ is replaced by
         $d^m(i,j)$ where $m > 1$. Show that the minimal spanning tree for the new
         network is the same as for the old. Using the result that

$$\lim_{m \to \infty} (a_1^m + a_2^m + \ldots + a_k^m) = (\max_{1 \leqslant i \leqslant k} (a_i))^m$$

show that the minimal spanning tree is also the solution for the problem
'Find the spanning tree whose longest arc is as short as possible'.

(2.4)   Use one of the connectedness algorithms to find whether the graph whose adjacency matrix is:

$$\begin{bmatrix} 0 & 0 & 1 & 0 & 0 \\ 0 & 0 & 0 & 1 & 0 \\ 1 & 0 & 0 & 0 & 1 \\ 0 & 1 & 0 & 0 & 1 \\ 0 & 0 & 1 & 1 & 0 \end{bmatrix}$$

       is connected or not.

(2.5)   Modify the computer program for Kruskal's algorithm to give the maximal spanning tree.

(2.6)   Some graphs are only connected because of the existence of one specific arc, such as the arc (5,6) in the graph below (Fig. 2.10). How can the connectedness algorithms be used to identify such arcs?

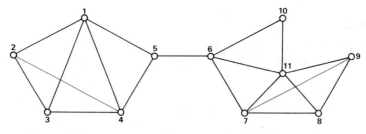

Fig. 2.10.

(2.7)   In the graph of exercise (2.6), how many paths with five arcs or less are there between nodes 5 and 6, allowing for repetitions? How can the appropriate connectedness algorithm be modified to answer this question?

3

Optimal Paths in Networks

This chapter examines some of the methods which find paths through a network which possess optimal properties. The general heading for these methods will be "shortest path problems", although it is not always necessary to be concerned with measuring the length of a path; the same methods can be used for paths with costs associated with them, in which case one might search for the cheapest path, or for paths with risks associated with them, when the objective could be to find the safest. But, for simplicity, the discussion will refer to shortest paths, and to distances. First, an examination is made of the simplest problem under the general heading, of finding the shortest path between two particular nodes of the network; then the shortest path from a given node to all other nodes in the network, and a generalisation of this to the problem of finding the shortest path between every pair of nodes in the network; finally a method is given for a related problem, of the second shortest path between a pair of nodes.

3.1 THE SHORTEST PATH PROBLEM

Suppose there is a distance $d(i,j)$ associated with the arc (i,j) in the network N. Then everyday experience assures one that the distance from node i to node k via node j will be the sum $d(i,j) + d(j,k)$. This result will be used to build up the distance between any two nodes s (the start of the path) and t (the terminus of the path) in N, given a path through N of the form $(s,i_1), (i_1, i_2), \ldots, (i_p,t)$. The distance will be $d(s,i_1) + d(i_1,i_2) + \ldots + d(i_p,t)$. In most networks, there will be many paths from s to t, and the aim will be to identify the shortest of these, and to calculate its length. There are several methods which have been produced for dealing with this problem, but the most widely used, and one of the most efficient, is due to Dijkstra [14].

3.1.1 Dijkstra's algorithm

In this method, the first step is to ensure that there is a distance associated with every pair of nodes in the network. This distance will be the arc length if there is an arc between the nodes (and the shortest arc length if there are several arcs between a pair of nodes), zero for the distance from a node to itself, and infinity for the distance between any pair of nodes which are not linked by an arc. (In practice, a very large number is substituted for infinity within computer programs.) It is not necessary for the distance from i to j to be equal to that from j to i (that is, $d(i, j)$ need not equal $d(j, i)$). This allows for one-way streets, or diversions, in which the distance travelled depends on the direction of travel.

Dijkstra's method assigns a label to every node in the network. This label is the distance to that node from the start (s) along the shortest path found thus far. The label can be in one of two states: it may be a permanent label, in which case the distance found is along the shortest of all paths, or it may be temporary, corresponding to some uncertainty as to whether the path found is the shortest of all. The method gradually changes temporary labels into permanent ones. (In some text-books, the two states of the labels are called coloured and un-coloured, or permanent and tentative. The precise wording does not matter, so long as it is clear which labels are in which state.) Given a set of nodes with temporary labels, the aim is to try and make these labels smaller by finding paths to these nodes using the shortest paths to permanently labelled nodes, followed by an arc from a node with a permanent label. Once this has been done, the node with the smallest temporary label is selected, and its label made permanent. This process is repeated until the terminus (t) has been assigned a permanent label, which must happen eventually, since every time the algorithm is used, one less temporary label is left, and so the number of nodes with temporary labels decreases to zero.

Formally the Dijkstra algorithm is as follows:

step 0   Assign a temporary label $l(i) = \infty$ to all nodes $i \neq s$; set $l(s) = 0$ and set $p = s$. Make $l(s)$ permanent. (p is the last node to be given a permanent label).

step 1   For each node i with a temporary label, redefine $l(i)$ to be the smaller of $l(i)$ and $l(p) + d(p, i)$. Find the node i with the smallest temporary label, set p equal to this i, and make the label $l(p)$ permanent.

step 2   If node t has a temporary label, then repeat step 1.
          Otherwise, t has a permanent label, and this corresponds to the length of the shortest path from s to t through the network. Stop.

3.1.2 Worked Example
Consider the application of the algorithm to the problem of finding the shortest path from s to t in the network shown in Fig. 3.1.

Distance matrix

$$
\begin{array}{c}
 & \begin{array}{cccccc} s & 1 & 2 & 3 & 4 & t \end{array} \\
\begin{array}{c} s \\ 1 \\ 2 \\ 3 \\ 4 \\ t \end{array}
\left[
\begin{array}{cccccc}
0 & 29 & 57 & \infty & 106 & \infty \\
29 & 0 & \infty & 90 & 97 & \infty \\
57 & \infty & 0 & 83 & 49 & \infty \\
\infty & 90 & 83 & 0 & \infty & 35 \\
106 & 97 & 49 & \infty & 0 & 78 \\
\infty & \infty & \infty & 35 & 78 & 0
\end{array}
\right]
\end{array}
$$

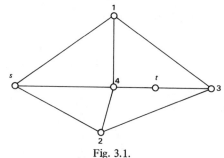

Fig. 3.1.

step 0   Assign labels as in the table below:
set $p = s$, make $l(s)$ permanent.

	s	1	2	3	4	t
$l(\ )$	0	∞	∞	∞	∞	∞
permanent?	yes	no	no	no	no	no

step 1   Redefine labels as below:

$l(1) = \min(\infty, 0 + 29) = 29$
$l(2) = \min(\infty, 0 + 57) = 57$
$l(3) = \min(\infty, 0 + \infty) = \infty$
$l(4) = \min(\infty, 0 + 106) = 106$
$l(t) = \min(\infty, 0 + \infty) = \infty$
set $p = 1$, make $l(1)$ permanent.

node	s	1	2	3	4	t
$l(\ )$	0	29	57	∞	106	∞
permanent?	yes	yes	no	no	no	no

step 2   Repeat step 1
step 1   Redefine labels as below:
$l(2) = \min(57, 29 + \infty) = 57$
$l(3) = \min(\infty, 29 + 90) = 119$
$l(4) = \min(106, 29 + 97) = 106$
$l(t) = \min(\infty, 29 + \infty) = \infty$
set $p = 2$, make $l(2)$ permanent.

node	s	1	2	3	4	t
$l(\ )$	0	29	57	119	106	∞
permanent?	yes	yes	yes	no	no	no

step 2   Repeat step 1
step 1   Redefine labels as below:

$l(3) = \min(119, 57 + 83) = 119$
$l(4) = \min(106, 57 + 49) = 106$
$l(t) = \min(\infty, 57 + \infty) = \infty$
set $p = 4$, make $l(4)$ permanent.

node	s	1	2	3	4	t
$l(\ )$	0	29	57	119	106	∞
permanent?	yes	yes	yes	no	yes	no

step 2   Repeat step 1
step 1   Redefine labels as below:

$l(3) = \min(119, 106 + \infty) = 119$
$l(t) = \min(\infty, 106 + 78) = 184$
set $p = 3$, make $l(3)$ permanent.

node	s	1	2	3	4	t
$l(\ )$	0	29	57	119	106	184
permanent?	yes	yes	yes	yes	yes	no

step 2   Repeat step 1
step 1   Redefine labels as below:

$l(t) = \min(184, 119 + 35) = 154$
set $p = t$, make $l(t)$ permanent.

node	s	1	2	3	4	t
$l(\ )$	0	29	57	119	106	154
permanent?	yes	yes	yes	yes	yes	yes

step 2   Stop. The shortest path through the network from node s to node t has
          length 154 units.

3.1.3 Notes

Several points emerge from this example. Firstly, in this case, permanent labels
have been attached to every node, and so the shortest path from node s to all
the other nodes has been found. This need not always happen, but, if one were
concerned with finding the shortest path from node s to all other nodes, the
Dijkstra algorithm could be modified by changing the stopping rule in step 2
to read:

step 2   If any node has a temporary label then repeat step 1.

Otherwise, all nodes have permanent labels, and we have found the shortest path from s to all nodes. Stop.

Secondly, since the original network had only undirected arcs, the shortest path from s to t must be the same as the shortest path from t to s. If there were any directed arcs, this result would no longer apply in all cases.

Thirdly, a potential weakness of the basic algorithm is visible; there is no indication of the composition of the optimal path. The only information is the length of the path. To deduce the nodes that are visited along the optimal path, one can return to the working above, and observe that the value 154 wasestablished by considering the permanent label $l(3)$, so that arc $(3,t)$ was the final arc in the optimal path; $l(3)$ was itself established by considering the label $l(1)$, so that $(1,3)$ must lie on the optimal path, and finally, $(s,1)$ must be a member of the optimal path. Hence the shortest route from s to t comprises arcs $(s,1)$, $(1,3)$, $(3,t)$.

To circumvent this problem, one may either create a second label on each node i as the algorithm is executed, indicating the preceding node in the path whose length is $l(i)$, and changing this when a shorter path is found, or use the approach which has been outlined informally above, and create the path from the permanent labels at the end of the algorithm. The former is slightly more effecient. However, for the sake of simplicity, the second method will be used here, as this only requires the introduction of an extra step to the algorithm, and the deletion of the 'Stop' instruction from step 2. (A return to algorithms which need two-part labels is made in later chapters.) The extra step will find a predecessor node $r(i)$ for each node i, as being the last node to be visited on the shortest path from s to i. Then one is able to build up the shortest path in reverse order, as being:

$$(r(t),t),(r(r(t)),r(t)),(r(r(r(t))),r(r(t))), \ldots (s,r(r\ldots)))$$

The new step of the algorithm will be:

step 3   For each permanently labelled node j other than s of the network, define $r(j) = i$ where $l(j) = l(i) + d(i,j)$ and $i \neq j$. (If there are several values i which satisfy this, select one arbitrarily, in which case there will be more than one path of minimal length to j.) Stop.

In the worked example:

$r(1) = s$
$r(2) = s$
$r(3) = 1$
$r(4) = 1$ (or 2)
$r(t) = 3$

and so the path is $(3,t)$, $(1,3)$, $(s,1)$.

3.1.4 Implementation

Labels such as those required by this algorithm are usually stored most conveniently as vectors, with each component of the vector corresponding to a node. The distances between nodes can be readily stored as a matrix, although in some cases it may be less wasteful of the storage space to store this information as a vector with pointers to the nodes being referred to. (This is most likely to be the case where there are realtively few arcs in the network, so most of the distances in the matrix would be infinite.) The indicator as to whether a label is temporary or permanent can be implemented in one of several ways; in a computer language with logical (Boolean) variables, then an array of these, set to true or false, could be used. Otherwise, a vector of integers set to 0 or 1 is possible, although both these methods would lead to an inefficient use of the store, depending on the compiler and language in use. If storage is at a premium, it is possible, at the expense of extra computing time, to store the state of a label as a part of the label itself. Thus permanent labels would have a value equal to −1 times their true value, and temporary ones would have positively valued labels. The calculation would identify the state by the sign of the label, and calculations would make use of the absolute value of each label.

Step 1 requires the scanning of a vector of labels, which can be readily performed using a loop, and the calculation of the smaller of two quantities. In the PASCAL program, this has been done using a function MIN, written as part of the program, although many languages offer this function as a standard one.

Step 2 requires a test of the state of a label: in BASIC this is followed by a GOTO statement, while in the PASCAL version, the whole of step 1 has been included in a REPEAT . . . UNTIL loop.

Since PASCAL offers logical arrays, these have been used: in BASIC it has been necessary to use an array of 0s and 1s.

```
program prog3p1(input,output);
  const t5 = 20;
        inf= 9999;
  var t0,t4,t6,t7,t8,i,j,n,p,s,t : integer;
      q : array[1..t5] of boolean;
      o,l,r : array[1..t5] of integer;
      d : array[1..t5,1..t5] of integer;
  function min(a,b:integer): integer;
  begin
     if (a<b) then min := a else min := b
  end;   { end of min }
  begin
  { Dijkstra's algorithm for the shortest
##    path between two given nodes
in a network.   The variables are as follows:
  q is a vector which is true if the
##    corresponding node has a permanent label
  o is a vector used to store the ordered
##    path at the end of the algorithm
  l is the vector of temporary/permanent labels
  r is the vector of predecessor nodes
```

```
  d is the matrix of arc lengths
  s is the start node, t is the terminal node,
##   p is the latest node
  to be given a permanent label, n is the
##   total number of nodes, and
  i and j are loop counters;  t0...t8 are
##   working area
  }
  writeln('      SHORTEST   PATH   ALGORITHM
##   (DIJKSTRA)');
  writeln;
  write('How many nodes?   ');
  repeat
    readln(n)
  until (n>0);
  if (n>t5) then
  begin
    writeln(' The program has been set up for
##   networks with at most',t5);
    writeln('nodes.  If you want to use
##   larger networks, then please edit');
    writeln('the program with a larger value
##   for t5 in the const section.')
  end
  else
  begin
    for i := 1 to n do
    begin
      for j:= 1 to n do d[i,j] := inf;
      o[i] := 0;
      d[i,i] := 0
    end;
    write('Are all the arcs undirected?
##   Answer 1 for yes, 0 for no  ');
    readln(t0);
    writeln('Enter arcs in the form:  start
##   node  finish node  distance');
    writeln('Enter 0  0  0  to finish ');
    t6 := 0;
    repeat
      t6 := t6 + 1;
      repeat
        write('Arc number ',t6:3,'  ');
        readln(i,j,t7);
        if ((i>n)or(j>n)) then
writeln('Node number too large:  try again');
        if ((i<0)or(j<0)) then
writeln('Node number too small:  try again')
until ((i>=0)and(i<=n)and(j>=0)and(j<=n));
      if (i<>0) then
        begin d[i,j] := t7; if (t0=1) then d[j,
##   i] := t7 end
    until (i=0);
    writeln('The matrix has been input as
##   follows: ');
    for i := 1 to n do
    begin for j := 1 to n do write(d[i,j]);
      writeln
    end;
    repeat { this repeat corresponds to the
##   desire for further shortest paths }
writeln('Which shortest path do you want?');
      repeat
writeln('Type start node  terminus node.');
        read(s,t);
        if ((s>n)or(t>n)) then
```

```
          writeln('Node number too high:   try again ');
              if ((s<1)or(t<1)) then
                  writeln('Node number too small:
##        try again ');
              if (s=t) then
                  writeln('Start and terminus are
##        the same:   try again ');
          until
##        ((s<>t)and(s>=1)and(t>=1)and(s<=n)and(t<=n));
          {
                      ******
                      step 0
                      ******
          }
          for i := 1 to n do
          begin
            l[i] := inf;
            q[i] := false
          end;
          q[s] := true;
          l[s] := 0;
          p := s;
          {
                      ******
                      step 1
                      ******
          }
          repeat   { this repeat corresponds to
##        the test in step 2 }
          for i := 1 to n do
          if (not q[i]) then
              l[i] := min(l[i],l[p]+d[p,i]);
          p := 0;
          t4 := inf ;
          for i := 1 to n do
          if ((not q[i]) and (t4 >= l[i])) then
          begin
            p := i;
            t4 := l[i]
          end;
          q[p] := true;
          {
                      ******
                      step 2
                      ******
          }
        until q[t];
        { a path to node t has been found }
        {
                      ******
                      step 3
                      ******
        }
        for j := 1 to n do
        if (q[j]) then
        for i := 1 to n do
        if ((i<>j) and (l[j] = l[i]+d[i,j])) then
        begin
          r[j] := i;
          if (d[i,j] >= inf) then
              writeln('No path possible.  A link
##        has been forced along',i,' to',j)
        end;
        t8 := t5;
        o[t8] := t;
        i := r[t];
```

```
while (i<>s) do
begin
  t8 := t8-1;
  o[t8] := i;
  i := r[i]
end;
writeln(' Path found is ');
write(s:3);
for i := t8 to t5 do
  write('-',o[i]:3);
writeln;
writeln(' This has length ',l[t]);
writeln(' Do you want any more paths in
## this network?   ');
write('  Enter 1 for yes, 0 for no:  ');
read(t0);
until (t0<>1)
end
end.
```

```
90 REM PROG3P2
100 DIM Q(20),L(20),O(20),R(20),D(20,20)
110 T5=20
120 REM DIJKSTRA ALGORITHM FOR THE
    SHORTEST PATH
130 REM VECTORS USED ARE Q : 0 FOR
    TEMPORARY LABEL
140 REM                       1 FOR
    PERMANENT LABEL
150 REM                    L : VALUE OF LABEL
160 REM                    R :
    PREDECESSOR NODE
170 REM                    0 : ORDERED PATH
180 REM MATRIX             D : DISTANCES
    ALONG ARCS
190 REM VARIABLES USED FOR WORKING ARE T0-T8
200 FOR I=1 TO T5
210 FOR J= 1 TO T5
220  D(I,J) = 9999
230 NEXT J
240 O(I)=0
250 D(I,I) = 0
260 NEXT I
270 PRINT "          SHORTEST PATH
    ALGORITHM (DIJKSTRA) "
280 PRINT
290 PRINT "HOW MANY NODES?   ";
300 INPUT N
310 IF N>T5 THEN 1290
320 PRINT "ARE ALL THE ARCS TWO-WAY?
    ANSWER 1 FOR YES, 0 FOR NO ";
330 INPUT T0
340 PRINT "ENTER ARCS IN THE FORM:
    START NODE,FINISH NODE,";
350 PRINT "DISTANCE.  ENTER 0,0,0 TO
    FINISH "
360 T6=0
370 T6=T6+1
380 PRINT "ARC NUMBER ";T6;
390 INPUT I,J,T7
400 IF I<=0 THEN 470
410 IF I>N THEN 1350
420 IF J>N THEN 1350
```

```
430 D(I,J)=T7
440 IF T0=0 THEN 370
450 D(J,I)=T7
460 GOTO 370
470 PRINT "THE MATRIX HAS BEEN INPUT AS
    BELOW"
480 FOR I=1 TO N
490 FOR J = 1 TO N
500 PRINT D(I,J);
510 NEXT J
520 PRINT
530 NEXT I
540 PRINT "WHICH SHORTEST PATH DO YOU WANT?"
550 PRINT "TYPE START NODE, TERMINUS
    NODE   ";
560 INPUT S,T
570 IF S>N THEN 1370


580 IF T>N THEN 1370
590 IF S<1 THEN 1390
600 IF T<1 THEN 1390
610 IF S=T THEN 1410
620 REM                         ======
630 REM                         STEP 0
640 REM                         ======
650 FOR I = 1 TO N
660 L(I) = 9999
670 Q(I) = 0
680 NEXT I
690 Q(S) = 1
700 L(S) = 0
710 P=S
720 REM                         ======
730 REM                         STEP 1
740 REM                         ======
750 FOR I = 1 TO N
760 IF Q(I) = 1 THEN 800
770 T1 = L(P) + D(P,I)
780 IF L(I) < T1 THEN 800
790 L(I) = T1
800 NEXT I
810 P=0
820 T2 = 9999
830 FOR I = 1 TO N
840 IF Q(I)=1 THEN 880
850 IF T2 < L(I) THEN 880
860 P=I
870 T2=L(I)
880 NEXT I
890 Q(P)=1
900 REM                         ======
910 REM                         STEP 2
920 REM                         ======
930 IF Q(T)=0 THEN 730
940 REM PATH FOUND
950 REM                         ======
960 REM                         STEP 3
970 REM                         ======
980 FOR J= 1 TO N
990 IF Q(J) = 0 THEN 1100
1000 FOR I = 1 TO N
1010 IF I = J THEN 1090
1020 T3 = L(I) + D(I,J)
```

```
1030 IF L(J) < T3 THEN 1090
1040 R(J) = I
1050 IF D(I,J) < 9998 THEN 1100
1060 PRINT "NO PATH POSSIBLE:    A LINK
     HAS BEEN FORCED ALONG ";
1070 PRINT I;" TO ";J
1080 GOTO 1280
1090 NEXT I
1100 NEXT J
1110 T8=T5
1120 O(T8)=T
1130 I=R(T)
1140 T8=T8-1
1150 O(T8)=I
1160 IF I=S THEN 1190



1170 I=R(I)
1180 GOTO 1140
1190 PRINT "PATH FOUND IS ";
1200 FOR I=T8 TO T5
1210   PRINT "-";O(I);
1220 NEXT I
1230 PRINT "THIS HAS LENGTH ";L(T)
1240 PRINT "DO YOU WANT ANY MORE PATHS
     IN THIS NETWORK?"
1250 PRINT "ENTER 1 FOR YES, 0 FOR NO   ";
1260 INPUT T0
1270 IF T0=1 THEN 540
1280 STOP
1290 PRINT "THE PROGRAM HAS BEEN SET UP
     FOR NETWORKS"
1300 PRINT "WITH AT MOST 20 NODES.   IF
     YOU WANT TO USE"
1310 PRINT "A LARGER NETWORK, PLEASE
     REDIMENSION THE"
1320 PRINT "MATRIX AND ARRAYS IN LINE
     100 AND CHANGE"
1330 PRINT "THE VALUE OF T5 IN LINE 110"
1340 STOP
1350 PRINT "NODE NUMBER TOO HIGH,
     PLEASE RETYPE"
1360 GOTO 380
1370 PRINT "NODE NUMBER TOO HIGH,
     PLEASE RETYPE"
1380 GOTO 540
1390 PRINT "NODE NUMBER TOO LOW,
     PLEASE RETYPE"
1400 GOTO 540
1410 PRINT "START AND TERMINUS ARE
     IDENTICAL,   PLEASE RETYPE"
1420 GOTO 540
```

3.2 GENERALISED SHORTEST PATHS

An important assumption in the Dijkstra algorithm is that all the arc lengths are positive or zero. When considering distances, this assumption is a reasonable one. However, if the numbers associated with the arcs are costs, then it is possible for some of these to be negative. For instance, a freight haulage contractor might assign a cost to an arc on which his lorries travelled empty, and a profit (a negative cost) to arcs on which the lorries carried loads. (In practice, of course, the contractor will face a more complex decision problem than simply choosing a route to minimise cost; for instance, he will have to decide whether to look for loads in certain circumstances, and will be constrained by legislation regarding loads and travel times.) An algorithm due to Ford [16] allows negative arc costs to be included in the network in problems of shortest paths, provided that there are no circuits around which the net cost is negative. (If such circuits exist, then the cost of any path which is incident on a node of such a circuit can be reduced to any desired figure, simply by including the circuit as many times as is necessary.) This means that no two-way arc can have a negative cost.

The introduction of negative costs on arcs leads to an algorithm which is very similar to the Dijkstra algorithm, but which no longer requires the two-state labels. It necessarily finds the shortest route from the start to every node in the network.

3.2.1 Ford's algorithm
In this algorithm, labels are assigned to each of the nodes, which represent the shortest path found to the node thus far. On each iteration of the algorithm, a search is made for a label which can be reduced, using the same approach as in the Dijkstra method. The algorithm terminates when no labels can be reduced. Formally, one may state the algorithm thus:

step 0   Assign a label $l(i) = \infty$ to each node i in the network, and set $l(s) = 0$.

step 1   For each node j, test whether there is an arc (i,j) such that $l(i) + d(i,j) < l(j)$. If there is such an arc, go to step 2. If no such arc exists for any node, stop.

step 2   Change $l(j)$ to $l(i) + d(i,j)$. Repeat step 1.

On termination, $l(t)$ is the length of the shortest path from s to t, as before; the actual path may be identified, either within the algorithm, or using a final stage, as in the step 3 of the Dijkstra algorithm above.

Since the algorithm fails when there is a circuit with a negative net cost, it is essential to take precautions when using the method to prevent the occurrence of such circuits. The first precaution will be to forbid two-way arcs with negative costs, but larger circuits cannot so easily be recognised. However, their existence can be proved if the label on any node is changed an excessive number of times. The algorithm will be terminated when the number of changes to any label reaches n, the number of nodes in the network. In order that this may be implemented, a count must be made of the number of times that step 2 is executed for each node j. The limit n is set because it is possible (though very unlikely) that the label on a node could be changed $n-1$ times in a network where there were no circuits with net negative cost. (This would correspond to a shortest path which passed through every node in the network, and which was created by increasing a path from one arc to $n-1$ arcs.)

3.2.2 Worked Example

A businessman plans to travel between two towns s and t as shown in the network below (Fig. 3.2). Along certain arcs, he can conduct business (sales, seminars etc.) which yield an effective profit, and on others, there is a cost associated with his journey. The costs associated with each arc are shown, and profits are represented as negative costs. What route should he take in order to minimise his costs?

Distance matrix

	s	1	2	3	t
s	0	+12	+10	∞	∞
1	∞	0	−3	−1	−5
2	∞	∞	0	+4	−2
3	∞	∞	−4	0	+10
t	∞	∞	∞	∞	0

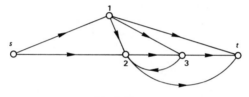

Fig. 3.2.

step 0   Assign labels:

node	s	1	2	3	t
$l(\ )$	0	∞	∞	∞	∞

step 1   Test node s : $l(i) + d(i,s) \geqslant \neq l(s)$ for all i.
Test node 1 : $l(s) + d(s,1) < l(1)$. Do step 2.

node	s	1	2	3	t	
$l(\ )$	0	12	∞	∞	∞	Do step 1.

step 1   Test node s : $l(i) + d(i,s) \geqslant \neq l(s)$ for all i.
Test node 1 : $l(i) + d(i,1) \geqslant \neq l(1)$ for all i.
Test node 2: $l(s) + d(s,2) < l(2)$. Do step 2.

step 2   Change label on node 2.

node	s	1	2	3	t
$l(\ )$	0	12	10	∞	∞

step 1   Test node s : $l(i) + d(i,s) \geqslant \neq l(s)$ for all i.
Test node 1 : $l(i) + d(i,1) \geqslant \neq l(1)$ for all i.
Test node 2: $l(1) + d(1,2) < l(2)$. Do step 2.

step 2   Change label on node 2:

node	s	1	2	3	t
$l(\ )$	0	12	9	∞	∞

                                        Do step 1.

step 1   Test node s: $l(i) + d(i,s) \geqslant \neq l(s)$ for all i.
        Test node 1: $l(i) + d(i,1) \geqslant \neq l(1)$ for all i.
        Test node 2: $l(i) + d(i,2) \geqslant \neq l(2)$ for all i.
        Test node 3: $l(1) + d(1,3) < l(3)$. Do step 2.

step 2   Change label on node 3:

node	s	1	2	3	t
$l(\ )$	0	12	9	11	∞

                                        Do step 1.

step 1   Test node s: $l(i) + d(i,s) \geqslant \neq l(s)$ for all i.
        Test node 1: $l(i) + d(i,1) \geqslant \neq l(1)$ for all i.
        Test node 2: $l(3) + d(3,2) < l(2)$. Do step 2.

step 2   Change label on node 2:

node	s	1	2	3	t
$l(\ )$	0	12	7	11	∞

                                     Do step 1.

step 1   Test node s: $l(i) + d(i,2) \geqslant \neq l(s)$ for all i.
        Test node 1: $l(i) + d(i,1) \geqslant \neq l(1)$ for all i.
        Test node 2: $l(i) + d(i,2) \geqslant \neq l(2)$ for all i.
        Test node 3: $l(i) + d(i,3) \geqslant \neq l(3)$ for all i.
        Test node t: $l(1) + d(1,t) < l(t)$. Do step 2.

step 2   Change label on node t:

node	s	1	2	3	t
$l(\ )$	0	12	7	11	7

Do step 1.

step 1   Test node s: $l(i) + d(i,s) \geqslant \neq l(s)$ for all i.
        Test node 1: $l(i) + d(i,1) \geqslant \neq l(1)$ for all i.
        Test node 2: $l(i) + d(i,2) \geqslant \neq l(2)$ for all i.
        Test node 3: $l(i) + d(i,3) \geqslant \neq l(3)$ for all i.
        Test node t: $l(2) + d(2,t) < l(t)$. Do step 2.

step 2   Change label on node t:

node	s	1	2	3	t
$l(\ )$	0	12	7	11	5

Do step 1.

step 1   Test node s: $l(i) + d(i, s) \geqslant \neq l(s)$ for all i.
Test node 1: $l(i) + d(i,1) \geqslant \neq l(1)$ for all i.
Test node 2: $l(i) + d(i,2) \geqslant \neq l(2)$ for all i.
Test node 3: $l(i) + d(i,3) \geqslant \neq l(3)$ for all i.
Test node t: $l(i) + d(i,t) \geqslant \neq l(t)$ for all i.
No change to any label: stop.

The least expensive path from node s to node t costs 5 units.

3.2.3 Implementation

The programs for this algorithm are very similar to those used for the Dijkstra algorithm, althouth there is no longer any need for testing the state of the labels. Instead, a count is kept of the number of times each label is changed, and this is stored as an integer vector. When too many changes to any label occur, a message is printed out, and the program halts, since it is not possible to find any shortest path in the network with certainty. (This program does not identify the circuit(s) which are causing this difficulty. If it is desired to identify it(them), then this can be done using the out-of-kilter algorithm (Chapter 4), setting the upper bounds on each arc equal to 1, the lower bounds to 0; the optimal flow for this network will have a flow of one unit in every arc of the circuit with the most negative net cost.)

```
program prog3p3(input,output);
 const t5 = 20;
        inf= 9999;
  var t4,t6,t7,t8,i,j,n,s,t:   integer;
       c,o,l,r : array[1..t5] of integer;
       d : array[1..t5,1..t5] of integer;
       loop,allok : boolean;
 begin
    {
       Ford's algorithm for the shortest path
##     between a given node and all
     other nodes in the network.   The variables
##     are as follows:
     t4..t8 are integers used for working, and
##     input/output
     allok  is a boolean variable used to test
##     for changes to the labels:   it
          is true when there have been changes
##     to the labels and false if a
          change has occurred.
     loop   is a boolean variable used to test
##     for loops of net negative cost
     and becomes true when such a loop is found.
     o   is a vector used to store an ordered
##     path at the end of the algorithm
     l   is the vector of labels
     r   is the vector of predecessor nodes
     d   is the matrix of arc lengths
     c   is the vector counting the number of
##     times a node has been labelled
     s   is the start node
     t   is the terminus node
```

```
   n  is the total number of nodes
   i and j are used as loop counters
   }
   writeln('  Shortest path algorithm (Method
##    of Ford)');
   writeln;
   write('  How many nodes?  ');
   repeat
     readln(n)
   until (n>0);
   if (n>t5) then
   begin
     writeln('  The program has been set up
##    for networks with at most ',t5);
     writeln('nodes.   If you want to use a
##    larger network, then the constant');
     writeln('t5 should be increased, and the
##    program recompiled.')
   end
   else
   begin
     for i := 1 to n do
     begin
       for j := 1 to n do
         d[i,j] := inf;
       o[i] := 0;
   program prog3_3(input,output);
     end;
     writeln('  Enter arcs in the form:  start
##    node  finish node  distance ');
     writeln('  Enter 0 0 0 to finish');
     t6 := 0;
     repeat
       repeat
         t6 := t6 +1;
           write('Arc number ',t6:3,'  ');
           readln(i,j,t7);
           if ((i>n)or(j>n)) then writeln('Node
##    number too high:  try again');
           if ((i<0)or(j<0)) then writeln('Node
##    number too low:  try again')
   until (((i>=0)and(j>=0)and(i<=n)and(j<=n));
       if (i<>0) then d[i,j] := t7
     until (i=0);
     writeln('  The matrix has been input as
##    follows:  ');
     for i :=  1 to n do
     begin
       for j :=  1 to n do  write(d[i,j]);
       writeln
     end;
     repeat { this repeat corresponds to the
##    desire for further shortest paths }
       writeln('  Which shortest path do you
##    require? ');
       repeat
   writeln('  Type start node  terminus node ');
         readln(s,t);
           if ((s<1)or(t<1)) then writeln('Node
##    number too high:  try again');
           if ((s>n)or(t>n)) then writeln('Node
##    number too low:  try again');
           if (s=t) then writeln('Start and
##    terminus are identical:  try again')
         until
##    ((s<>t)and(s>=1)and(t>=1)and(s<=n)and(t<=n));
       {
```

```
              ******
              Step 0
              ******
  }
      for i :=  1 to n do
      begin
        c[i] := 0;
        l[i] := inf
      end;
      l[s] := 0;
  {
              ******
              Step 1
              ******
  }
      loop := false;
      repeat
        allok := true;
        for i :=  1 to n do
        for j :=  1 to n do
          if (l[i] > l[j] + d[j,i]) then
          begin
  {
              ******
              Step 2
              ******
  }
            allok := false;
            l[i] := l[j] + d[j,i];
            c[i] := c[i] +1;
            if (c[i] = n) then loop := true
          end
      until (allok or loop);
  {
              ******
              Step 3
              ******
  }
      if (loop) then
        writeln('There is a circuit with net
  ##  negative cost')
      else
      begin
        for i :=  1 to n do
        for j :=  1 to n do
  if ((i<>j)and(l[j] = l[i] + d[i,j])) then
          begin
            r[j] := i;
            if (d[i,j] >= inf) then
              writeln('No path possible: a
  ##  link has been forced along',i,' to',j)
          end;
        t8 := t5;
        o[t8] := t;
        i := r[t];
        while (i<>s) do
        begin
          t8 := t8 - 1;
          o[t8] := i;
          i := r[i]
        end;
        writeln('Path found is ');
        write(s:3);
  for i := t8 to t5 do write('-',o[i]:3);
  writeln; writeln('This has length ',l[t]);
        end;
```

```
        if (loop) then t4 := 0
        else
        begin
           writeln('Do you want any further
##      paths in this network?');
write('Enter 1 for yes, 0 for no   :');
           read(t4)
        end
     until (t4 <> 1)
     end
   end.
```

```
90 REM PROG3P4
100 DIM L(20),O(20),R(20),C(20),D(20,20)
110 T5=20
120 REM FORD ALGORITM FOR THE SHORTEST PATH
130 REM VECTORS USED ARE L : VALUE OF LABEL
140 REM                   R :
    PREDECESSOR NODE
150 REM                   O : ORDERED PATH
160 REM                   C : NUMBER OF
    TIMES EACH NODE HAS BEEN
170 REM                       LABELLED
180 REM MATRIX            D : DISTANCES
    ALONG ARCS
190 REM VARIABLES USED FOR WORKING ARE T0-T8
200 FOR I=1 TO T5
210 FOR J= 1 TO T5
220   D(I,J) = 9999
230 NEXT J
240 O(I)=0
250 D(I,I) = 0
260 NEXT I
270 PRINT "         SHORTEST PATH
    ALGORITHM (FORD)"
280 PRINT
290 PRINT "HOW MANY NODES?   ";
300 INPUT N
310 IF N>T5 THEN 1180
320 PRINT "ENTER ARCS IN THE FORM:
    START NODE,FINISH NODE,";
330 PRINT "DISTANCE.   ENTER 0,0,0 TO
    FINISH "
340 T6=0
350 T6=T6+1
360 PRINT "ARC NUMBER ";T6;
370 INPUT I,J,T7
380 IF I<=0 THEN 430
390 IF I>N THEN 1240
400 IF J>N THEN 1240
410 D(I,J)=T7
420 GOTO 350
430 PRINT "THE MATRIX HAS BEEN INPUT AS
    BELOW"
440 FOR I=1 TO N
450 FOR J = 1 TO N
460 PRINT D(I,J);
470 NEXT J
480 PRINT
490 NEXT I
500 PRINT "WHICH SHORTEST PATH DO YOU WANT?"
```

```
510 PRINT "TYPE START NODE, TERMINUS
    NODE   ";
520 INPUT S,T
530 IF S>N THEN 1260
540 IF T>N THEN 1260
550 IF S<1 THEN 1280
560 IF T<1 THEN 1280
570 IF S=T THEN 1300
580 REM                        ======
590 REM                        STEP 0
600 REM                        ======


610 FOR I = 1 TO N
620 L(I) = 9999
630 C(I)=0
640 NEXT I
650 L(S)=0
660 REM                        ======
670 REM                        STEP 1
680 REM                        ======
690 FOR I = 1 TO N
700 FOR J=1 TO N
710 T1 = L(J) + D(J,I)
720 IF L(I) > T1 THEN 760
730 NEXT J
740 NEXT I
750 GOTO 850
760 REM                        ======
770 REM                        STEP 2
780 REM                        ======
790 L(I) = T1
800 C(I) = C(I) + 1
810 IF C(I) < N THEN 690
820 REM THERE IS A CIRCUIT
830 PRINT "THERE IS A CIRCUIT WITH NET
    NEGATIVE COST"
840 STOP
850 REM                        ======
860 REM                        STEP 3
870 REM                        ======
880 FOR J = 1 TO N
890   FOR I = 1 TO N
900   IF I = J THEN 980
910   T3 = L(I) + D(I,J)
920   IF L(J) < T3 THEN 980
930   R(J) = I
940   IF D(I,J) < 9998 THEN 990
950   PRINT "NO PATH POSSIBLE;   A LINK
    HAS BEEN FORCED ALONG ";
960   PRINT I;" TO ";J
970   GOTO 1170
980   NEXT I
990   NEXT J
1000 T8=T5
1010 O(T8)=T
1020 I=R(T)
1030 T8=T8-1
1040 O(T8)=I
1050 IF I=S THEN 1080
1060 I=R(I)
1070 GOTO 1030
1080 PRINT "PATH FOUND IS ";
1090 FOR I=T8 TO T5
```

```
1100  PRINT "-";O(I);
1110 NEXT I
1120 PRINT "THIS HAS LENGTH ";L(T)
1130 PRINT "DO YOU WANT ANY MORE PATHS
     IN THIS NETWORK?"
1140 PRINT "ENTER 1 FOR YES, 0 FOR NO  ";
1150 INPUT T0
1160 IF T0=1 THEN 500
1170 STOP


1180 PRINT "THE PROGRAM HAS BEEN SET UP
     FOR NETWORKS"
1190 PRINT "WITH AT MOST 20 NODES,  IF
     YOU WANT TO USE"
1200 PRINT "A LARGER NETWORK, PLEASE
     REDIMENSION THE"
1210 PRINT "MATRIX AND ARRAYS IN LINE
     100 AND CHANGE"
1220 PRINT "THE VALUE OF T5 IN LINE 110"
1230 STOP
1240 PRINT "NODE NUMBER TOO HIGH,
     PLEASE RETYPE"
1250 GOTO 360
1260 PRINT "NODE NUMBER TOO HIGH,
     PLEASE RETYPE"
1270 GOTO 500
1280 PRINT "NODE NUMBER TOO LOW,
     PLEASE RETYPE"
1290 GOTO 500
1300 PRINT "START AND TERMINUS ARE
     IDENTICAL,  PLEASE RETYPE"
1310 GOTO 500
```

3.3 AN ALGORITHM TO FIND ALL SHORTEST PATHS

Both the algorithms described above will find the shortest paths from a given start node s to all other nodes. If they were to be used n times, once with each node of the network as the start node, then the shortest path between every pair of nodes in the network would be found. This could be summarised in a distance chart of the form familiar to motorists and users of atlases. (Construction of a road distance chart requires a large network, using far more information than distances between centres of population.) In place of such a laborious procedure, use is made of an algorithm for finding all shortest paths in a network. This method is due to Floyd [15]. (A similar, related, method is due to Dantzig [13].)

3.3.1 Floyd's algorithm

In Floyd's method, the nodes are numbered from 1 to n. The algorithm builds up the shortest paths between pairs of nodes, first finding the shortest paths which are either direct or which use node 1 as an intermediate node, then the shortest paths which are direct or which use nodes 1 and/or 2 as intermediate nodes, and continues in this fashion until the shortest paths which are either direct or which use some or all of the nodes 1 to n as intermediate nodes have been found. This then provides the matrix of shortest paths. The distances found at any stage of the algorithm, say when the nodes 1 to k may be used as intermediate nodes, are stored in an $n*n$ matrix D_k, with elements $d_k(i,j)$. ($d_k(i,j)$ is the shortest distance from i to j via some or all of the nodes $1 \ldots k$.) Then the elements $d_{k+1}(i,j)$ of the next matrix will either be $d_k(i,j)$ (if there is no advantage in including node $k+1$ in the path) or $d_k(i,k+1) + d_k(k+1,j)$ (if including node $k+1$ is beneficial), and the smaller of these will be chosen. The arc lengths of the network will form the matrix D_0, and the final matrix – the distance chart – will be D_n.

Stated formally, the Floyd algorithm is:

step 0   Create the $n*n$ matrix D_0 whose elements are:
$d_0(i,j) = d(i,j)$ (the length of arc (i,j), if this exists),
$= 0$ (if $i = j$),
$= \infty$ (if no arc (i,j) exists).
Set $k = 0$.

step 1   Define the $n*n$ matrix D_{k+1} with elements $d_{k+1}(i,j) = \min(d_k(i,j),$
$d_k(i,k+1) + d_k(k+1,j))$.

step 2   Increase k by 1. If $k = n$, stop, otherwise return to step 1.

When the algorithm stops, it will have found the shortest path lengths between every pair of nodes in the network. The nodes which make up this path can be found, as in the Dijkstra and Ford algorithms, either by including an evaluation of the preceding node within the algorithm, or by calculating this node by means of the original distance matrix once the algorithm has found D_n. (The predecessor of node t in the shortest path from s to t will be that node k which satisfies $d_n(s,t) = d_n(s,k) + d(k,t)$.) In computational terms, the former method of finding the nodes on the path is normally quicker, but necessitates a larger program. The modified algorithm, incorporating an evaluation of the path, will be:

step 0   Create the $n*n$ matrix D_0 whose elements are:
$d_0(i,j) = d(i,j)$ (the length of arc (i,j), if this exists),
$= 0$ (if $i = j$),
$= \infty$ (if no arc (i,j) exists).
Create the $n*n$ matrix P with elements:
$p(i,j) = i$.
Set $k = 0$.

step 1   Define the $n*n$ matrix D_{k+1} with elements $d_{k+1}(i,j) = \min(d_k(i,j),$
$d_k(i, k+1) + d_k(k+1,j))$.
If the latter value is chosen, then change $p(i,j)$ to be $p(k+1,j)$.

step 2   Increase k by 1. If $k = n$, stop, otherwise return to step 1.

When this algorithm stops, the matrix P contains integers in the range $1 \ldots n$; the value of $p(i,j)$ is the last node to be visited on the shortest path from i to j, and so the path can built up (backwards) using arcs $(p(i,j),j)$, $(p(i,p(i,j))$, $p(i,j)) \ldots (i,p(i,p(i, \ldots (i,j) \ldots)$.

3.3.2 Worked Example
Consider the modified algorithm applied to the network shown below (Fig. 3.3):

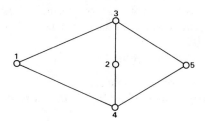

Fig. 3.3.

step 0 $D_0 = \begin{bmatrix} 0 & \infty & 97 & 90 & \infty \\ \infty & 0 & 83 & 49 & \infty \\ 97 & 83 & 0 & \infty & 35 \\ 90 & 49 & \infty & 0 & 78 \\ \infty & \infty & 35 & 78 & 0 \end{bmatrix}$ $P = \begin{bmatrix} 1 & 1 & 1 & 1 & 1 \\ 2 & 2 & 2 & 2 & 2 \\ 3 & 3 & 3 & 3 & 3 \\ 4 & 4 & 4 & 4 & 4 \\ 5 & 5 & 5 & 5 & 5 \end{bmatrix}$

      $k = 0$

step 1 $d_1(1,1) = \min(0, 0 + 0) = 0$ $p(1,1) = 1$
          $d_1(1,2) = \min(\infty, 0 + \infty) = \infty$ $p(1,2) = 1$
          $d_1(1,3) = \min(97, 0 + 97) = 97$ $p(1,3) = 1$
          $d_1(1,4) = \min(90, 0 + 90) = 90$ $p(1,4) = 1$
          $d_1(1,5) = \min(\infty, 0 + \infty) = \infty$ $p(1,5) = 1$

          $d_1(2,1) = \min(\infty, \infty + 0) = \infty$ $p(2,1) = 2$
          $d_1(2,2) = \min(0, 0 + 0) = 0$ $p(2,2) = 2$
          $d_1(2,3) = \min(83, \infty + 97) = 83$ $p(2,3) = 2$
          $d_1(2,4) = \min(49, \infty + 90) = 49$ $p(2,4) = 2$
          $d_1(2,5) = \min(\infty, \infty + \infty) = \infty$ $p(2,5) = 2$

          $d_1(3,1) = \min(97, 97 + 0) = 97$ $p(3,1) = 3$
          $d_1(3,2) = \min(83, 97 + \infty) = 83$ $p(3,2) = 3$
          $d_1(3,3) = \min(0, \infty + \infty) = 0$ $p(3,3) = 3$
          $d_1(3,4) = \min(\infty, 97 + 90) = 187$ $p(3,4) = 1$
          $d_1(3,5) = \min(35, 97 + \infty) = 35$ $p(3,5) = 3$

$$d_1(4,1) = \min(90, 90 + 0) = 90 \qquad p(4,1) = 4$$
$$d_1(4,2) = \min(49, 90 + \infty) = 49 \qquad p(4,2) = 4$$
$$d_1(4,3) = \min(\infty, 90 + 97) = 187 \qquad p(4,3) = 1$$
$$d_1(4,4) = \min(0, 90 + 90) = 0 \qquad p(4,4) = 4$$
$$d_1(4,5) = \min(78, 90 + \infty) = 78 \qquad p(4,5) = 4$$

$$d_1(5,1) = \min(\infty, \infty + 0) = \infty \qquad p(5,1) = 5$$
$$d_1(5,2) = \min(\infty, \infty + \infty) = \infty \qquad p(5,2) = 5$$
$$d_1(5,3) = \min(35, \infty + 97) = 35 \qquad p(5,3) = 5$$
$$d_1(5,4) = \min(78, \infty + 90) = 78 \qquad p(5,4) = 5$$
$$d_1(5,5) = \min(0, \infty + \infty) = 0 \qquad p(5,5) = 5$$

So $D_1 = \begin{bmatrix} 0 & \infty & 97 & 90 & \infty \\ \infty & 0 & 83 & 49 & \infty \\ 97 & 83 & 0 & 187 & 35 \\ 90 & 49 & 187 & 0 & 78 \\ \infty & \infty & 35 & 78 & 0 \end{bmatrix} \quad P = \begin{bmatrix} 1 & 1 & 1 & 1 & 1 \\ 2 & 2 & 2 & 2 & 2 \\ 3 & 3 & 3 & 1 & 3 \\ 4 & 4 & 1 & 4 & 4 \\ 5 & 5 & 5 & 5 & 5 \end{bmatrix}$

step 2 $k = 1$ (not equal to 5); do step 1.

step 1 Using the same method as before, we calculate D_2 and P to be:

$$D_2 = \begin{bmatrix} 0 & \infty & 97 & 90 & \infty \\ \infty & 0 & 83 & 49 & \infty \\ 97 & 83 & 0 & 132 & 35 \\ 90 & 49 & 132 & 0 & 78 \\ \infty & \infty & 35 & 78 & 0 \end{bmatrix} \quad P = \begin{bmatrix} 1 & 1 & 1 & 1 & 1 \\ 2 & 2 & 2 & 2 & 2 \\ 3 & 3 & 3 & 2 & 3 \\ 4 & 4 & 2 & 4 & 4 \\ 5 & 5 & 5 & 5 & 5 \end{bmatrix}$$

step 2 $k = 2$ (not equal to 5); do step 1.

step 1 Using the same method as before, we calculate D_3 and P to be:

$$D_3 = \begin{bmatrix} 0 & 180 & 97 & 90 & 132 \\ 180 & 0 & 83 & 49 & 118 \\ 97 & 83 & 0 & 132 & 35 \\ 90 & 49 & 132 & 0 & 78 \\ 132 & 118 & 35 & 78 & 0 \end{bmatrix} \quad P = \begin{bmatrix} 1 & 3 & 1 & 1 & 3 \\ 3 & 2 & 2 & 2 & 3 \\ 3 & 3 & 3 & 2 & 3 \\ 4 & 4 & 2 & 4 & 4 \\ 3 & 3 & 5 & 5 & 5 \end{bmatrix}$$

step 2 $k = 3$ (not equal to 5); do step 1.

step 1 Using the same method as before, we calculate D_4 and P to be:

$$D_4 = \begin{bmatrix} 0 & 139 & 97 & 90 & 132 \\ 139 & 0 & 83 & 49 & 118 \\ 97 & 83 & 0 & 132 & 35 \\ 90 & 49 & 132 & 0 & 78 \\ 132 & 118 & 35 & 78 & 0 \end{bmatrix} \quad P = \begin{bmatrix} 1 & 4 & 1 & 1 & 3 \\ 4 & 2 & 2 & 2 & 3 \\ 3 & 3 & 3 & 2 & 3 \\ 4 & 4 & 2 & 4 & 4 \\ 3 & 3 & 5 & 5 & 5 \end{bmatrix}$$

step 2 $k = 4$ (not equal to 5); do step 1.

step 1   Using the same method as before, we calculate D_5 and P to be:

$$D_5 = \begin{bmatrix} 0 & 139 & 97 & 90 & 132 \\ 139 & 0 & 83 & 49 & 118 \\ 97 & 83 & 0 & 113 & 35 \\ 90 & 49 & 113 & 0 & 78 \\ 132 & 118 & 35 & 78 & 0 \end{bmatrix} \quad P = \begin{bmatrix} 1 & 4 & 1 & 1 & 3 \\ 4 & 2 & 2 & 2 & 3 \\ 3 & 3 & 3 & 5 & 3 \\ 4 & 4 & 5 & 4 & 4 \\ 3 & 3 & 5 & 5 & 5 \end{bmatrix}$$

step 2 $k = 5$: stop.

3.3.3 Note

It is apparent from the example that this algorithm is tedious to use for hand calculation. It can be speeded up in step 2 by a considerable amount, by not using the recurrence relation for the cases $d_{k+1}(i, i)$, which will always be zero, and the cases $d_{k+1}(k+1, i)$ and $d_{k+1}(i, k+1)$, since these will always equal the corresponding elements of D_k. By not using the recurrence relationship in these cases, we reduce the number of times that a comparision of values has to be made from $n * n$ to $(n-1) * (n-2)$ per iteration, which saves a total of $(3 * n * n - 2 * n)$ comparisons in the whole of the algorithm.

3.3.4 Implementation

The algorithm has been presented in terms of $n * n$ matrices, and these are the most efficient means for storing the data. However, it is not necessary to use $n + 1$ matrices for the data $D_0, D_1, \ldots D_n$, since the algorithm only makes use of the preceding matrix on any iteration. So, in step 1, we retain only two matrices, an 'old matrix' and a 'new matrix' (which correspond respectively to D_k and D_{k+1}). In step 2, we copy the 'new matrix' onto the 'old matrix', in readiness for repeating step 1. It is possible, although neither program does this, to avoid such copying altogether, and to use a pointer to identify which of the two matrices was most recently altered. The BASIC and PASCAL programs are very similar. This is a consequence of the relatively straightforward calculations which are necessary, and for which neither language offers features which the other lacks.

```
program prog3p5.p(input, output);
   const maxnodes = 20;
         inf = 9999;
   var   t0, t1, t2, t3, t4, t6, t7, i, j, k, n : integer;
         d, e, p : array[1..maxnodes,
##       1..maxnodes] of integer;
   begin
   { Floyd's algorithm
   This algorithm finds all shortest paths in
##    a network with undirected
   nodes, yielding both the lengths of all
##    paths and the preceding node
   in the path (which allows the whole path to
```

```
##    be built up)
variables used:
n       : number of nodes
i,j     : the ends of the arc being read in
d       : matrix of shortest paths found thus
##    far
e       : matrix created from d, including
##    further reductions in path length
p       : matrix of preceding nodes
k,t0,t1,t2,t3,t4,t6,t7 : loop counters
}
writeln('     Floyd algorithm for all
##    shortest paths in a network ');
writeln;
write('  How many nodes?');
repeat
  readln(n)
until (n>0);
if (n>maxnodes) then
begin
  writeln('     The program has been set up
##    for networks with at most',maxnodes);
  writeln('nodes,   If you want to use a
##    larger network, then the constant');
  writeln('maxnodes should be increased,
##    and the program recompiled,')
end
else
begin
  for t3 := 1 to n do
  begin
    for t4 := 1 to n do  d[t3,t4] := inf;
    d[t3,t3] := 0
  end;
  writeln('Are all the arcs two-way?
##    Enter 1 for yes, 0 for no');
  readln(t0);
  writeln('Enter arcs in the form:  start
##    node finish node distance');
  writeln('Enter 0 0 0 to finish');
  t6 := 0;
  repeat
    t6 := t6 + 1;
    repeat
      write('Arc number',t6:3,'  ');
      readln(i,j,t7);
      if ((i>n)or(j>n)) then writeln('Node
##    number too high:  try again');
      if ((i<0)or(j<0)) then writeln('Node
##    number too low:  try again');
      if ((i=j)and(i>0)) then
        writeln('Start and finish should be
##    different:  try again')
    until (((i in [1..n])and(j in
##    [1..n])and(i<>j))or(i=0));
    if (i<>0) then
    begin d[i,j] := t7;  if (t0=1) then d[j,
##    i] := t7 end
  until (i=0);
  writeln('The matrix has been input as
##    below:');
  for t3 := 1 to n do
  begin
    for t4 := 1 to n do
    begin
      write(d[t3,t4]);
```

```
            p[t3,t4] := t3
        end;
        writeln
    end;
{                   ******
                    STEP 0
                    ******
}
    k := 0;
{                   ******
                    STEP 1
                    ******
}
    repeat
        t2 := k+1;
        for t3 := 1 to n do
        for t4 := 1 to n do
        begin
            if ((t3=t2)or(t4=t2)or(t3=t4)) then
            e[t3,t4] := d[t3,t4]
            else
            begin
                t1 := d[t3,t2]+d[t2,t4];
                if (t1>d[t3,t4]) then e[t3,t4] :=
##          d[t3,t4] else
                begin
                    e[t3,t4] := t1;
                    p[t3,t4] := p[t2,t4]
                end
            end
        end;
{                   ******
                    STEP 2
                    ******
}
        k := k+1;
        for t3 := 1 to n do
for t4 := 1 to n do d[t3,t4] := e[t3,t4];
    until (k=n);
    writeln('The matrix of shortest distances
##      is ');
        for t3 := 1 to n do
        begin
            for t4 := 1 to n do write(d[t3,t4]);
            writeln
        end;
        writeln;
writeln('The matrix of predecessors is ');
        for t3 := 1 to n do
        begin
            for t4 := 1 to n do write(p[t3,t4]);
            writeln
        end
    end
    end.
```

```
90 REM PROG3P6
100 DIM D(20,20),E(20,20),P(20,20)
110 T5=20
120 REM FLOYD ALGORITHM FOR ALL
    SHORTEST PATHS IN A NETWORK
130 REM MATRICES USED ARE D : INITIAL
    DISTANCE MATRIX
140 REM                          : AND "OLD"
    MATRIX
150 REM                      E : "NEW" MATRIX
160 REM                      P :
    PREDECESSOR NODES
170 REM VARIABLES USED FOR WORKING ARE T0-T8
180 FOR I=1 TO T5
190 FOR J= 1 TO T5
200  D(I,J) = 9999
210 NEXT J
220 D(I,I) = 0
230 NEXT I
240 PRINT "         SHORTEST PATH
    ALGORITHM (FLOYD) "
250 PRINT
260 PRINT "HOW MANY NODES?  ";
270 INPUT N
280 IF N>T5 THEN 1000
290 PRINT "ARE ALL THE ARCS TWO-WAY?
    ANSWER 1 FOR YES, 0 FOR NO ";
300 INPUT T0
310 PRINT "ENTER ARCS IN THE FORM:
    START NODE,FINISH NODE,";
320 PRINT "DISTANCE,  ENTER 0,0,0 TO
    FINISH "
330 T6=0
340 T6=T6+1
350 PRINT "ARC NUMBER ";T6;
360 INPUT I,J,T7
370 IF I<=0 THEN 440
380 IF I>N THEN 1060
390 IF J>N THEN 1060
400 D(I,J)=T7
410 IF T0=0 THEN 340
420 D(J,I)=T7
430 GOTO 340
440 PRINT "THE MATRIX HAS BEEN INPUT AS
    BELOW"
450 FOR I=1 TO N
460 FOR J = 1 TO N
470 PRINT D(I,J);
480 P(I,J) = I
490 NEXT J
500 PRINT
510 NEXT I
520 REM                    ======
530 REM                    STEP 0
540 REM                    ======
550 REM D AND P HAVE ALREADY BEEN  SET UP
560 K=0
570 REM                    ======
580 REM                    STEP 1
590 REM                    ======
600 T2 = K+1
610 FOR I = 1 TO N
620 FOR J = 1 TO N
630 IF I=T2 THEN 680
640 IF J=T2 THEN 680
650 IF I=J THEN 680
```

```
660   T1 = D(I,T2) + D(T2,J)
670   IF T1 < D(I,J) THEN 700
680   E(I,J) = D(I,J)
690   GOTO 720
700   E(I,J) = T1
710   P(I,J) = P(T2,J)
720   NEXT J
730   NEXT I
740   REM                    ======
750   REM                    STEP 2
760   REM                    ======
770   K =K + 1
780   IF K = N THEN 850
790   FOR I = 1 TO N
800   FOR J = 1 TO N
810    D(I,J) = E(I,J)
820   NEXT J
830   NEXT I
840   GOTO 570
850   PRINT "THE MATRIX OF SHORTEST PATHS
      IS :"
860   FOR I = 1 TO N
870   FOR J = 1 TO N
880   PRINT  E(I,J);
890   NEXT J
900   PRINT
910   NEXT I
920   PRINT " THE MATRIX OF PREDECESSORS IS :"
930   FOR I = 1 TO N
940   FOR J = 1 TO N
950   PRINT P(I,J);
960   NEXT J
970   PRINT
980   NEXT I
990   STOP
1000  PRINT "THE PROGRAM HAS BEEN SET UP
      FOR NETWORKS"
1010  PRINT "WITH AT MOST 20 NODES.   IF
      YOU WANT TO USE"
1020  PRINT "A LARGER NETWORK, PLEASE
      REDIMENSION THE"
1030  PRINT "MATRICES  IN  LINE  100
      AND  CHANGE"
1040  PRINT "THE VALUE OF T5 IN LINE 110"
1050  STOP
1060  PRINT "NODE NUMBER TOO HIGH,
      PLEASE RETYPE"
1070  GOTO 350
```

3.4 OTHER SHORTEST PATH PROBLEMS

3.4.1 The second shortest path (Pollack's algorithm)

It is sometimes desirable to know not just the shortest path through a network, but also the superiority of this over any of its rivals. To discover this, it is necessary to identify the length (and possibly the path) of the second shortest path through the network. There are several methods for identifying this, and the one to be presented here is that due to Pollack [17]. This is the simplest of the methods, although it is not necessarily the most efficient. It relies on the (inherently obvious) result that the second shortest path differs from the shortest path in at least one arc, and will be the shortest path which satisfies this extra condition. The method proceeds by finding the shortest paths through a series of networks, which are modified replicas of the original network, but in the ith network in the series, the length of the ith arc of the shortest path is temporarily set to infinity. The shortest of all the shortest paths through these modified networks will be the second shortest path through the original network.

It may happen that the path found by this method has the same length as the original shortest path. This corresponds to the possibility that there were two equally short paths in the network, and that which has been identified as the shortest was chosen in an arbitrary way. This may be satisfactory for one's purposes, but situations may arise in which the user wants the length of the path to be different from the shortest overall path. The method can be extended to provide this result, since the path which satisfies this criterion will be different in at least one arc from all the paths which tied for the title 'shortest path'. This can be found by setting pairs of arcs to infinite length, and finding the shortest of all paths through the modified network. Pollack's algorithm can be formally stated as:

step 0   Find (using the Dijkstra algorithm) the shortest path through the network. Number the arcs in this path 1 to m, where arc l is from $i(l-1)$ to $i(l)$ and $i(0) = s$, $i(m) = t$. Set $k = 1$, $q = \infty$, $r = 0$.

step 1   Set $d(i(k-1), i(k)) = \infty$ temporarily. Find the shortest path through the resulting network. If this has length less than q, then
(i) set q equal to the length of the path
(ii) store the path
(iii) set $r = k$.

step 2   If $k = m$ stop. Otherwise set $k = k + 1$ and repeat step 1.

The algorithm terminates with the second shortest path stored, and its length stored as q. This path corresponds to the elimination of the rth arc from the original shortest path.

It will be seen that this algorithm uses the Dijkstra algorithm as part of two of its steps. This is fairly common feature of algorithms, in that we may use one as a building block for another, and it is a further remainder (as if one were needed) of the versatility of this approach to problem solving.

Example of the Pollack algorithm
Consider once again the network in section 3.1.2:

	s	1	2	3	4	t
s	0	29	57	∞	106	∞
1	29	0	∞	90	97	∞
2	57	∞	0	83	49	∞
3	∞	90	83	0	∞	35
4	106	97	49	∞	0	78
t	∞	∞	∞	35	78	0

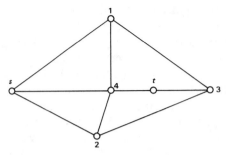

Fig. 3.4.

step 0   The shortest path from s to t is $(s,1)$, $(1,3)$, $(3,t)$, with length 154.
Set $k = 1$, $q = \infty$ and $r = 0$.

step 1   Set $d(s,1) = \infty$. In the modified network, the shortest path has length
175, and is $(s,2)$, $(2,3)$, $(3,t)$. Set $q = 175$, $r = 1$ and store this path.

step 2 $k = 2$. Repeat step 1.

step 1   Set $d(1,3) = \infty$ (and $d(s,1) = 29$). In the modified network, the shortest
path again has length 175 and is again $(s,2)$, $(2,3)$, $(3,t)$.

step 2 $k = 3$. Repeat step 1.

step 1   Set $d(3,t) = \infty$ (and $d(1,3) = 90$). In the modified network, the shortest
path has length 184 and is $(s,2)$, $(2,4)$, $(4,t)$.

step 2   Stop. The second shortest path has length 175 and corresponds to the
removal of arc 1 (or 2) from the original shortest path.

3.4.2 Implementation of Pollack's algorithm

The Pollack method involves relatively few changes to the basic Dijkstra algorithm,
which is used as a sub-routine (named DIJK in PASCAL) of the programs
presented here. It is necessary to introduce a number of extra variables, to store
the information about the paths found at each step, and to retain details of the
second shortest path.

    As in the Dijkstra programs, the shortest path is stored as a vector, with the
path being built up backwards from the right-most end of the vector. It is thus
necessary to index the arcs in the shortest path to correspond to this method of
storing the path, and so the counter for the arc (identified as k in the algorithm)
is initially set to the location in the vector corresponding to the first arc, and is
not set to 1.

```
program prog3p7(input,output);
   const maxnodes = 20;
         inf= 9999;
   type logarr = array[1..maxnodes] of boolean;
   intarr = array[1..maxnodes] of integer;
         dismat = array[1..maxnodes,
   ##    1..maxnodes] of integer;
   var dt,t0,t4,t6,t7,t8,t9,i,j,k,n,qa,ra,s,t:
```

```
##      integer:
        u,v,o,l,r : intarr:
        d : dismat;

  function min(a,b : integer):integer;
  begin
     if (a<b) then min := a else min := b
  end;

  procedure dijk(var r,l,o : intarr; var t8:
##      integer :
n,s,t : integer; var d : dismat);
  var i,j,p,t2 : integer;
      q : logarr;
  begin
     for i := 1 to n do
     begin
        l[i] := inf;
        q[i] := false
     end;
     q[s] := true;
     l[s] := 0;
     p := s;
     repeat
        for i := 1 to n do
        if (not q[i]) then
           l[i] := min(l[i],l[p]+d[p,i]);
        p := 0;
        t2 := inf;
        for i := 1 to n do
        if ((not q[i])and(t2 >=l[i])) then
        begin
           p := i;
           t2 :=l[i]
        end;
        q[p] := true;
     until q[t];
     for j := 1 to n do
     if (q[j]) then
     for i := 1 to n do
        if ((i<>j)and(l[j] = l[i] + d[i,j]))
##      then r[j] := i;
     t8 := maxnodes;
     o[t8] := t;
     i := r[t];
     while (i<>s) do
     begin
        t8 := t8-1;
        o[t8] := i;
        i := r[i]
     end
  end; {of dijk }

  begin
  (* Pollack's algorithm for the second
##      shortest path between two given nodes
in a network.  The variables are as follows:
   t0.. are variables used for working and
##      input/output
   q  is a vector which is true if the
##      corresponding node has a permanent label
   o  is a vector used to store the ordered
##      path at the end of the algorithm
   u  is a vector used to store the shortest
##      path
   v  is a vector used to store the best
```

```
##    alternative to u found so far
 l    is the vector of temporary/permanent
##    labels
 r    is the vector of predecessor nodes
 d    is the matrix of arc lengths
 s    is the start node, t is the terminus
 p    is the latest node to have been given a
##    permanent label
 n    is the total number of nodes
 i and j are loop counters
 k    is the loop counter when arcs are
##    sequentially set to infinity
 ga   is the length of the second shortest
##    path, corresponding to having
 set the   ra th  arc in the shortest path
##    to infinity
 *)
 writeln('  Second shortest path algorithm
##    (Pollack) ');
 writeln;
 write('  How many nodes?  ');
 repeat
   read(n)
 until (n>0);
 if (n>maxnodes) then
 begin
   writeln('The program has been set up for
##    networks with at most',maxnodes);
   writeln('nodes.  If you want to use a
##    larger network, then the constant');
   writeln('maxnodes at the head of the
##    program should be changed.')
 end
 else
 begin
   for i := 1 to n do
   begin
     for j := 1 to n do d[i,j] := inf;
     o[i] := 0;
     d[i,i] := 0
   end;
   write('Are all the arcs two-way?  Answer
##    1 for yes, 0 for no  ');
   readln(t0);
   writeln('Enter arcs in the form:  start
##    node  terminus node  distance');
   writeln('Enter 0 0 0 to stop');
   t6 := 0;
   repeat
     t6 := t6 + 1;
     repeat
       write('Arc number ',t6:3,'  ');
       read(i,j,t7);
       if ((i>n)or(j>n)) then
 writeln('Node number too high:  try again');
       if ((i<0)or(j<0)) then
 writeln('Node number too low:  try again');
       if (t7<0) then
         writeln('Distances must not be
##    negative:  try again')
     until ((i>=0)and(j>=0)
           and(i<=n)and(j<=n)and(t7>=0));
     if (i<>0) then d[i,j] := t7;
     if ((i<>0)and(t0=1)) then d[j,i] := t7
   until (i=0);
   writeln('The matrix has been input as
```

```
##    follows:');

   for i := 1 to n do

   begin
     for j := 1 to n do
       write(d[i,j]);
     writeln
   end;
   repeat { this corresponds to the wish for
##    more than one shortest path }
     writeln(' Which shortest path do you
##    want? ');
   repeat
writeln(' Type start node terminus node ');
     read(s,t);
     if ((s>n)or(t>n)) then
writeln('Node number too high;  try again');
     if ((s<1)or(t<1)) then
writeln('Node number too low;  try again');
     if (s=t) then
       writeln('Start and terminus are the
##    same.  Try again')
   until
##    ((s<>t)and(s>=1)and(s<=n)and(t>=1)and(t<=n));
   {
                ******
                STEP 0
                ******
   }
   dijk(r,l,o,t8,n,s,t,d);
   if (l[t]<inf) then
   begin
   writeln('     Shortest path found is');
     write(s:3);
     u[t8-1] := s;
     for i := t8 to maxnodes do
       begin
       write('-',o[i]:3);
       u[i] := o[i]
     end;
     writeln;
     writeln(' This has length ',l[t])
   end
   else writeln('There is no shortest path for
##    the given nodes');
     k := t8;
     t4 := t8;
     ga := inf;
     ra := 0;
   {
                ******
                STEP 1
                ******
   }
   repeat
     dt := d[u[k-1],u[k]];
     d[u[k-1],u[k]] := inf;
     dijk(r,l,o,t8,n,s,t,d);
     if (l[t] < ga) then
     begin
       ga := l[t];
       t9 := t8;
       ra := k-t4+1;
```

```
for i := t8 to maxnodes do v[i] := o[i];
    v[t8-1] := s
  end;
    d[u[k-1],u[k]] := dt;
 {

            ******
            STEP 2
            ******
 }
    k := k+1
  until (k>maxnodes);
  if (aa<inf) then
  begin
  writeln('Second shortest path found is');
  write(s:3);
  for i := t9 to maxnodes do
    write('-',v[i]:3);
  writeln;
  writeln('This has length ',aa);
  writeln('It corresponds to the deletion
##  of the ',ra:3,'th arc');
  writeln
  end
  else writeln('There is no second shortest
##  path for the given nodes');
  writeln('Do you want any more paths in
##  this network?');
  write('Enter 1 for yes, 0 for no:  ');
  read(t0)
  until (t0<>1)
  end
  end.
```

```
90  REM PROG3P8
100 REM PROG3P8
110 DIM U(20),V(20),O(20),L(20),R(20),
    D(20,20),Q(20)
120 M9 = 20
130 T1= 9999
140 REM POLLACK'S ALGORITHM FOR THE
    SECOND SHORTEST PATH
150 REM     BETWEEN TWO GIVEN NODES IN A
    NETWORK.
160 REM  THE VARIABLES ARE AS FOLLOWS:
170 REM T0.. ARE VARIABLES USED FOR
    WORKING AND INPUT/OUTPUT
180 REM Q  IS A VECTOR WHICH IS 1 IF
    THE CORRESPONDING
190 REM    NODE HAS A PERMANENT LABEL
200 REM O  IS A VECTOR USED TO STORE
    THE ORDERED PATH
210 REM    AT THE END OF THE ALGORITHM
220 REM U  IS A VECTOR USED TO STORE
    THE SHORTEST PATH
230 REM V  IS A VECTOR USED TO STORE
    THE BEST ALTERNATIVE TO U FOUND SO FAR
240 REM L  IS THE VECTOR OF
    TEMPORARY/PERMANENT LABELS
250 REM R  IS THE VECTOR OF PREDECESSOR
    NODES
260 REM D  IS THE MATRIX OF ARC LENGTHS
```

```
270 REM S   IS THE START NODE, T IS THE
    TERMINUS
280 REM P   IS THE LATEST NODE TO HAVE
    BEEN GIVEN A PERMANENT LABEL
290 REM N   IS THE TOTAL NUMBER OF NODES
300 REM I AND J ARE LOOP COUNTERS
310 REM K   IS THE LOOP COUNTER WHEN
    ARCS ARE SEQUENTIALLY SET TO INFINITY
320 REM Q0  IS THE LENGTH OF THE SECOND
    SHORTEST PATH,
330 REM     CORRESPONDING TO HAVING SET
    THE T3 TH ARC IN THE SHORTEST PATH
340 REM     TO INFINITY
350 REM S9 IS USED TO TEST FOR VALID DATA
360 PRINT "   SECOND SHORTEST PATH
    ALGORITHM (POLLACK)   "
370 PRINT
380 PRINT "HOW MANY NODES?   :";
390 INPUT N
400 IF N<=0 THEN 380
410 IF N<=M9 THEN 460
420  PRINT "THE PROGRAM HAS BEEN SET UP
    FOR NETWORKS WITH AT MOST",M9
430  PRINT "NODES.    IF YOU WANT TO USE
    A LARGER NETWORK, THEN THE "
440  PRINT "CONSTANT M9 AT THE HEAD OF
    THE PROGRAM SHOULD BE CHANGED. "
450 STOP
460 FOR I = 1 TO N
470 FOR J = 1 TO N
480  D(I,J) = T1
490 NEXT J
500 Q(I) = 0
510 D(I,I) = 0
520 NEXT I
530  PRINT "ARE ALL THE ARCS TWO-WAY?
    ANSWER 1 FOR YES, 0 FOR NO    ";
540 INPUT T0
550 PRINT "ENTER ARCS IN THE FORM:
    START NODE,FINISH NODE,LENGTH"
560 PRINT "ENTER 0,0,0 TO FINISH"
570  T6 = 0
580  T6 = T6 + 1
590  PRINT "ARC NUMBER ";T6;" ";
600  S9=0
610 INPUT I,J,T7
620  IF I>N GOSUB 1900
630  IF J>N GOSUB 1900
640  IF I<0 GOSUB 1930
650  IF J<0 GOSUB 1930
660  IF T7<0 GOSUB 1960
670  IF S9=1 THEN 590
680  IF I=0 THEN 730
690  D(I,J)=T7
700  IF T0=0 THEN 580
710  D(J,I)=T7
720  GOTO 580
730  PRINT "THE MATRIX HAS BEEN INPUT
    AS FOLLOWS: "
740  FOR I = 1 TO N
750    FOR J = 1 TO N
760    PRINT D(I,J);
770  NEXT J
780 PRINT
790 NEXT I
800 PRINT " WHICH SHORTEST PATH DO YOU
    WANT?  "
```

```
810 PRINT " TYPE START NODE,TERMINUS
    NODE   ";
820 INPUT S,T
830 S9=0
840 IF S>N GOSUB 1900
850 IF T>N GOSUB 1900
860  IF S<1 GOSUB 1930
870  IF T<1 GOSUB 1930
880     IF  S=T  GOSUB 1990
890 IF S9=1 THEN 800
900 REM           ******
910 REM             STEP 0
920 REM           ******
930 GOSUB 1520
940 IF L(T)<T1 THEN 970
950 PRINT "THERE IS NO SHORTEST PATH
    FOR THE GIVEN NODES"
960 STOP
970 PRINT "    SHORTEST PATH FOUND IS "
980  PRINT S;
990  U(T8-1) = S
1000 FOR I = T8 TO M9
1010   PRINT "-";O(I);
1020 U(I) = O(I)
1030 NEXT I
1040 PRINT
1050  PRINT " THIS HAS LENGTH ";L(T)
1060  K = T8
1070  T4 = T8
1080  Q0 = T1
1090  T3 = 0
1100 REM           ******
1110 REM             STEP 1
1120 REM           ******
1130 U0=U(K-1)
1140 U1=U(K)
1150 T2 =, D(U0,U1)
1160 D(U0,U1) = T1
1170 GOSUB 1520
1180 IF  L(T) > Q0   THEN 1270
1190   Q0 = L(T)
1200   T9 = T8
1210   T3 = K-T4+1
1220 FOR I = T8 TO M9
1230   V(I) = O(I)
1240 NEXT I
1250 U8=T8-1
1260 V(U8) = S
1270 U0 = U(K-1)
1280 U1 = U(K)
1290 D(U0,U1) = T2
1300 REM           ******
1310 REM             STEP 2
1320 REM           ******
1330 K = K+1
1340 IF K<=M9 THEN 1130
1350  IF Q0<T1 THEN 1380
1360  PRINT "THERE IS NO SECOND
    SHORTEST PATH FOR THE GIVEN NODES "
1370  GOTO 1470
1380  PRINT "SECOND SHORTEST PATH FOUND IS "
1390  PRINT S;
1400  FOR I = T9 TO M9
1410    PRINT "-";V(I);
1420  NEXT I
1430  PRINT
```

```
1440    PRINT "THIS HAS LENGTH ";Q0
1450    PRINT "IT CORRESPONDS TO THE
        DELETION OF THE ";T3;"TH ARC "
1460    PRINT
1470    PRINT "DO YOU WANT ANY MORE PATHS
        IN THIS NETWORK? "
1480    PRINT "ENTER 1 FOR YES, 0 FOR NO:    ";
1490  INPUT T0
1500  IF T0=1 THEN 800
1510  STOP
1520    FOR I = 1 TO N
1530      L(I) = T1
1540      Q(I) = 0
1550    NEXT I
1560    Q(S) = 1
1570    L(S) = 0
1580    P = S
1590    FOR I = 1 TO N
1600    IF Q(I)=1 THEN 1620
1610      L(I) = MIN(L(I),L(I),L(I),
        L(P)+D(P,I))
1620    NEXT I
1630    P = 0
1640    T2 = T1
1650    FOR I = 1 TO N
1660  IF Q(I)=1 THEN 1700
1670  IF T2<L(I) THEN 1700
1680    P = I
1690    T2 =L(I)
1700  NEXT I
1710    Q(P) = 1
1720    IF Q(T)=0 THEN 1590
1730    FOR J = 1 TO N
1740    IF  Q(J)=0 THEN 1800
1750    FOR I = 1 TO N
1760    IF I=J THEN 1790
1770    IF L(J)<>L(I)+D(I,J) THEN 1790
1780    R(J)=I
1790    NEXT I
1800    NEXT J
1810    T8 = M9
1820    O(T8) = T
1830    I = R(T)
1840  IF I=S THEN 1890
1850    T8 = T8-1
1860    O(T8) = I
1870    I = R(I)
1880    GOTO 1840
1890  RETURN
1900    PRINT "NODE NUMBER TOO HIGH:
        TRY AGAIN "
1910    S9=1
1920  RETURN
1930    PRINT "NODE NUMBER TOO LOW:    TRY
        AGAIN "
1940    S9=1
1950    RETURN
1960    PRINT "DISTANCES MUST NOT BE
        NEGATIVE:   TRY AGAIN "
1970    S9=1
1980    RETURN
1990    PRINT    "START AND TERMINUS ARE
        THE SAME,   TRY AGAIN "
2000    S9=1
2010    RETURN
```

3.4.3 The kth shortest path

If one were interested in paths other than the second shortest one, then use could be made of one of several methods for finding the paths which take a particular place in the rank. The general problem is to find all paths up to and including the kth shortest. To do this involves a generalisation of the idea of the minimum from a pair of numbers to a vector of integers; a detailed description of the main algorithms is provided by Minieka [3].

3.4.4 Paths with other properties

Many network problems can be cast into the mould of a shortest route problem, or can be treated as being equivalent to one. In all the algorithms examined in this chapter there are two essential features. First, for any path in the network, a particular property must be calculated or otherwise determined. Then, this property must be compared with the property associated with another path, and the path with the preferred property selected. In the case of shortest distance, distances have been added to yield the distance associated with a path, and the arithmetic operation of selecting a minimum has been used in order that the preferred path could be found. Alternatives abound. If there were to be a risk associated with each arc — such as the risk of robbery — and it were known that the risks along separate arcs were independent, then the multiplication rule for independent probabilities could be used to assign risks for separate paths, and the less risky one be chosen. Or, more lightheartedly, an individual might be able to select the prettiest path between two points, by gradually building up the alternatives, and eliminating those which had a prettier counterpart.

If all the arcs in the network are undirected, then the network can be represented by a scale model with string tied together at the nodes. (This returns to the idea of a network resembling a net.) The shortest path between any two nodes can be found by grasping the appropriate knots, and pulling the net apart. This method is not recommended for computer implementation!

EXERCISES

(3.1)   Use Dijkstra's algorithm to find the shortest path between nodes 1 and 6 of the network:

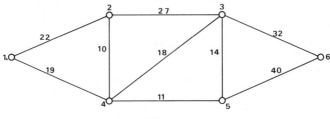

Fig. 3.5.

(3.2) What happens when Dijkstra's method is applied to the example given for Ford's algorithm?

(3.3) Find the shortest and second shortest paths between nodes 2 and 6 of the network in exercise (3.1).

(3.4) Give a formal algorithm, and demonstrate its application, for the problem of finding the safest route from node 1 to node 6 in the network below, where p_{ij} represents the probability of successfully using the arc. Probabilities in different arcs are independent.

$$\begin{bmatrix} 1.0 & 0.7 & 0.8 & 0.0 & 0.0 & 0.0 \\ 0.0 & 1.0 & 0.8 & 0.6 & 0.0 & 0.0 \\ 0.0 & 0.8 & 1.0 & 0.9 & 0.6 & 0.0 \\ 0.0 & 0.6 & 0.0 & 1.0 & 0.9 & 0.5 \\ 0.0 & 0.0 & 0.0 & 0.9 & 1.0 & 0.5 \\ 0.0 & 0.0 & 0.0 & 0.0 & 0.5 & 1.0 \end{bmatrix}$$

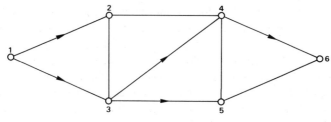

Fig. 3.6.

(3.5) Modify the computer program for Ford's algorithm so that all shortest paths between s and t are output.

(3.6) Repeat exercise (3.5), for Floyd's algorithm, assuming that s and t are selected by the user, once the algorithm has run.

(3.7) What is the second safest route through the network of exercise (3.4)?

4

Optimal Flow Algorithms

A water supply undertaking has a network of supply canals and pipes linking reservoirs, wells and pumping stations to its customers. Each of these has a limited capacity for supplying and transmitting water each day. In such a case it is reasonable to ask what is the largest daily flow which the system will meet. It may be the limit set by the links from the wells; or it may be the limit set by the pipes feeding the customers; or it may be neither. But what is the bound?

Besides the capacity associated with each part of the network, there may be a cost associated with each unit of flow which is sent along it — a cost of pumping, maintenance and other overheads. One might wish to ask how the customers could be supplied with a fixed amount of water each day at minimal cost? And, still thinking of the problems of the water supply company, it might be of benefit to know the extent of the disruption to supplies if a particular pipeline was broken or had to be repaired. In some such circumstance, a particular question could be asked: which pipeline is most crucial to the supply of water?

These problems are all related, as each deals with aspects of the flow of some material along links between points in what may be viewed as a network. The instance above is concerned with water flowing in various channels, but other materials could have been chosen equally well. The same, and other similar problems, arise in the study of the flow of traffic, of goods in a factory, of people and of telephone calls in a network of exchanges. All may be studied under the general heading "flow algorithms".

Broadly speaking, the idea of a flow is a common-sense one; some thing or things move from one point to another along a pre-determined route. The items being moved can be measured in some convenient units, and there are limits associated with the amount of flow which can take place along each part of the route in a given time. These different parts may be conveniently regarded as a network, in which there are arcs (where flows move from point to point) linking nodes (where flows join or are separated). In the simplest version of the problem, each arc (i,j) has an associated upper bound on flow, $u(i,j)$, and an actual flow, $x(i,j)$ which satisfies the inequalities $0 \leqslant x(i,j) \leqslant u(i,j)$. The value of $u(i,j)$ will be determined by the physical nature of the link, and will correspond to the bottleneck on that link — the place where flow is most limited. Whatever physical material it is which is being sent within the network, it is assumed that it may be divided into parts, to flow in different quantities in different arcs. This is obviously possible for most items, such as water and vehicles, provided that appropriate units are used for the measurement of these.

When flows in arcs meet at nodes, it is desirable that there is no gain or loss of material, except at the ends of the network. This gives rise to a constraint on the flows at each node that the sum of the flows into the node must equal the sum of the flows out of it. In terms of the flows, one has $\sum_i x(i,j) = \sum_k x(j,k)$ for node j.

The only exceptions are those nodes which have been (rather loosely) described as the ends of the network. These can be divided into two groups, those which act as sources for the flow, and those which act as destinations. In the simplest case, there is only one source, and only one destination; conventionally, these are known as the source and sink respectively, and are frequently referred to as nodes s and t. For this case, the problem to be considered for the network is that of finding the largest flow from s to t, using the arcs of the network, and subject to the flow constraints in the arcs and the conservation of flow at each node except source and sink.

Other problems, such as those with upper and lower bounds on flows, and those with several sources and/or sinks, are developments of this simplest one.

Underlying the algorithms for each is one concept of which regular use is made; this is the idea of a 'flow-augmenting chain', which is often referred to (loosely) as a 'flow-augmenting path'.

4.1 FLOW-AUGMENTING CHAINS

A **flow-augmenting chain** between a pair of nodes is a chain of arcs which connect the nodes, and which may be used to increase the flow from the one node to the other. In the case of flow in a network from source to sink, the chain will start at the source, s, and end at the sink, t. Two kinds of arcs may be used in such chains. First, there is the (obvious) kind, which will be an arc in the direction 's to t', and which has some spare capacity. In such an arc, the flow from s to t may be increased by increasing the flow in the arc. The second type of arc which may form a part of a flow-augmenting chain is in the opposite sense (that is, from t to s), and has non-zero flow in it. The flow from s to t may be increased, using such an arc, by reducing the flow in it. Since the flow from t to s is being lowered, there is a net gain in the flow from s to t. The first type of arcs are frequently known as forward arcs, and the second as (not surprisingly) reverse arcs. A flow augmenting chain will contain a mixture of these two types; in Fig. 4.1, there is a flow-augmenting chain made up wholly of forward arcs; in Fig. 4.2, one made up wholly of reverse arcs (which would be unusual); and in Fig. 4.3, there is a mixture of the two kinds.

Fig. 4.1 – A flow-augmenting chain with all forward arcs.

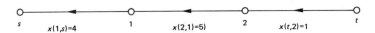

Fig. 4.2 – A flow-augmenting chain with all reverse arcs.

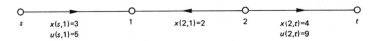

Fig. 4.3 – A flow-augmenting chain with both forward and reverse arcs.

The amount of change which can be achieved by use of a flow-augmenting chain is determined by the arc which is closest to its limit on flow. For forward arcs, this limit is the upper bound, and so the amount of extra flow is determined by the difference $u(i,j) - x(i,j)$; for reverse arcs, the limit is zero flow, and the extra flow is limited by the amount of reduction possible, that is, the flow $x(i,j)$. So, for a flow-augmenting chain whose links are $(s,i_1), (i_1,i_2), \ldots (i_r,i_{r+1}), \ldots (i_q,t)$ the volume of extra flow that is possible is found by calculating

$$\delta(i_r,i_{r+1}) = u(i_r,i_{r+1}) - x(i_r,i_{r+1}) \qquad \text{(for forward arcs)}$$

$$= x(i_{r+1},i_r) \qquad \text{(for reverse arcs)}$$

$$\text{capacity of chain} = \min_{\substack{\text{arcs in} \\ \text{chain}}}(\delta(i_r,i_{r+1})) \qquad \text{where } i_0 = s, \ i_{q+1} = t$$

So, for the chains shown in Figs. 4.1, 4.2 and 4.3, the possible extra flows are 3, 1 and 2 respectively.

The flows in the arcs in the chain are never changed so as to violate the constraints on them.

It is now possible to return to the original problem, that of finding the largest flow through a network from s to t.

4.2 AN ALGORITHM FOR MAXIMUM FLOW

There are two complementary parts in the algorithm for maximal flow from source to sink. In the first part, a search is made to find a flow-augmenting chain from s to t, using a labelling scheme. No special criteria are applied to such a chain; it may be able to carry one extra unit of flow, or hundreds; it may use one arc, or very many. Once it has been found, the second part of the algorithm is used to make the appropriate changes in the flows in each arc of the chain. Then, the first part of the algorithm is used again, to find another flow-augmenting chain, then the flow is changed along this, and so on. It is an iterative process which repeatedly seeks for chains and then changes flows in them until no further chains can be found. When this happens, then the flow from source to sink is at its largest possible value; the distribution of flows in the arcs of the network is one of the (perhaps many) possible allocations of flow which achieves this maximum total flow.

When finding a flow-augmenting chain, the algorithm makes use of a labelling scheme; each node is assigned a two-part label. The first component of the label indicates the source of possible extra flow, and the second, the amount of extra flow which can be sent along the chain from s to the node in question. The source of extra flow is the previous node in the chain, either the head of a forward arc, or the tail of a reverse arc. For the chain shown in Fig. 4.4, the labels are shown incomplete with the first part only calculated. Once the sink has been labelled, the desired chain is complete, and the second phase of the algorithm is used. This makes use of the second components of the labels, which determine the change in flow.

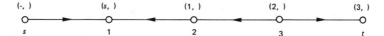

Fig. 4.4 – A flow-augmenting chain with partially-labelled nodes.

It would be possible, once a suitable chain had been found, to examine each of its constituent arcs in order that the extra flow be calculated. But it is more convenient to use the labels to indicate this, and the relevant parts are calculated recursively, as the labels are established. Suppose that node j is being labelled, and its immediate predecessor in the chain is node i, which is already labelled (a_i, b_i). (For obvious reasons, chains can only be extended from labelled to unlabelled nodes.) There are two cases to be examined:

(a) if the arc is a forward arc (i,j), then its potential capacity is $u(i,j) - x(i,j)$. The capacity of the chain from s to j will be the smaller of:
   (i)   the capacity of the chain from s to i, and
   (ii) the potential capacity of the arc (i,j).
   So, the label on node j is defined as $(i, \min(b_i, u(i,j) - x(i,j)))$.
(b) if the arc is a reverse arc (j,i), then its potential capacity for extra forward flow is $x(j,i)$. In the same way as in the first case, the capacity for extra flow of the chain from s to j is the smaller of:
   (i)   the capacity of the chain from s to i and
   (ii) this potential capacity.

Thus, the label on node j is $(i, \min(b_i, x(j,i)))$. (It is sometimes more convenient to identify a reverse arc by making the label $a_j = -i$, but this is not essential.)

To start this process of labelling, all nodes are initially unlabelled, and the source s is given a label of the form $(-, \infty)$, since its predecessor is undefined, and there are no limits on the amount of flow which can be sent along the chain from s to itself. Then nodes are labelled by finding arcs with one end labelled and the other not, and whose flows allow them to be used as arcs in a flow-augmenting chain. This entails either scanning the list of arcs for suitable ones, or examining a newly labelled node and the other nodes which can be reached from it. If there is a flow-augmenting chain then eventually node t will be labelled. Then, flow along this chain is increased by an amount b_t, and the chain itself is identified by working backwards from the sink to the source. Flow along the forward arcs in the chain is increased by b_t, while along reverse arcs the flow is decreased by the same amount. The second parts of the labels on nodes in the chain have no further value, except for those on the sink. Once this change is complete, then the labels are all removed, and the process repeated.

Eventually, no flow-augmenting chain will be found. This situation is recognisable when no more nodes can be given labels, and t is not labelled. There will be some nodes, forming a set S, with labels, in which the source node will be found. The complementary set $\bar{S}$ (containing t) will contain all the unlabelled nodes. To say that there is no node in $\bar{S}$ which can be labelled is equivalent to saying that all the arcs from S to $\bar{S}$ are carrying their maximum capacity of flow, and that all the arcs from $\bar{S}$ to S are carrying zero flow. This situation is illustrated in Fig. 4.5.

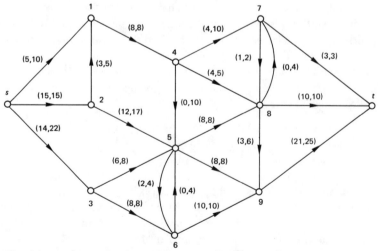

Fig. 4.5 – A network with labelled and unlabelled nodes. There is no flow-augmenting chain, notation (x,u) represents flow and bound.

$S = \{ s, 1, 2, 3, 5, 6 \}$ Labels:

$\bar{S} = \{ 4, 7, 8, 9, t \}$ on s $(-, \infty)$

                                                   1 $(s, 5)$

                                                   2 $(s, 1)$

                                                   3 $(s, 8)$

                                                   5 $(2, 5)$

                                                   6 $(5, 2)$

The algorithm, which was described by Ford and Fulkerson [18], can now be stated formally:

Algorithm for maximum flow in a network with upper limits on arcs.

step 0   (Initialisation). Give each arc a feasible flow, ensuring that flow is conserved at each node other than the source node s and the sink node t. (This may be done by assigning a zero flow to each arc.)

step 1   Label node s with the label $(-, \infty)$, and ensure that no other node is labelled.

step 2   Scan through the arcs until one (i,j) is found for which
either (a) node i is labelled *and* node j is not *and*

$$x(i,j) < u(i,j) \qquad \text{(a forward arc)}$$

or (b) node j is labelled *and* node i is not *and*

$$x(i,j) > 0 \qquad \text{(a reverse arc)}$$

If no such arc exists, go to step 5.

step 3   If (a) is true, then label node j with the two-part label (a_j, b_j) where $a_j = i$, $b_j = \min(b_i, u(i,j) - x(i,j))$. If (b) is true, then label node i with the two-part label (a_i, b_i) where $a_i = -j$, $b_i = \min(b_j, x(j,i))$. If node t is now labelled, do step 4, otherwise do step 2 again.

step 4   (A flow-augmenting chain has been found.) Increase the flow in the flow-augmenting chain by the amount b_t. If the node t is labelled (l, b_t), then increase the flow in the arc (l, t); if node t is labelled $(-l, b_t)$ then decrease the flow in (t, l). Now examine the label on node l, and repeat the same procedure, until the source is reached, always changing by b_t. Go to step 1.

step 5   The optimal flow has been found. Stop.

4.2.1 Worked Example

Suppose that the maximum flow from node 1 to node 4 in the network shown in Fig. 4.6 is being sought. Here, there is a network (N, A, U) where:

$$N = \{1,2,3,4\}$$
$$A = \{(1,2), (2,3), (3,4), (1,3), (2,4)\}$$
$$u(1,2) = 8; u(2,3) = 5; u(3,4) = 6; u(1,3) = 3; u(2,4) = 7.$$

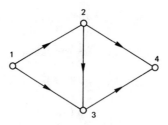

Fig. 4.6.

The arcs will be scanned in the (slightly contrived) order given in A. Using the algorithm, the progress at each step is:

step 0   Assign zero flows to all arcs.
step 1   Label node 1 $(-, \infty)$.
step 2   Arc $(1,2)$ is case (a).
step 3   Label node 2 $(1,\min(\infty, 8-0)) = (1,8)$.
step 2   Arc $(2,3)$ is case (a).
step 3   Label node 3 $(2,\min(8, 5-0)) = (2,5)$.
step 2   Arc $(3,4)$ is case (a).
step 3   Label node 4 $(3,\min(5, 6-0)) = (3,5)$, node 4 is now labelled.
step 4   Increase flow in the chain $(1,2)$, $(2,3)$, $(3,4)$ by 5
          so $x(3,4) = 0 + 5 = 5$
          $x(2,3) = 0 + 5 = 5$
          $x(1,2) = 0 + 5 = 5$.
step 1   Label node 1 $(-, \infty)$.
step 2   Arc $(1,2)$ is case (a).
step 3   Label node 2 $(1,\min(\infty,8-5)) = (1,3)$.
step 2   Arc $(1,3)$ is case (a).
step 3   Label node 3 $(1,\min(\infty, 3-0)) = (1,3)$.
step 2   Arc $(2,4)$ is case (a).
step 3   Label node 4 $(2,\min(3,7-0)) = (2,3)$, node 4 is now labelled.
step 4   Increase flow in the chain $(1,2)$, $(2,4)$ by 3
          so $x(2,4) = 0 + 3 = 3$
          $x(1,2) = 5 + 3 = 8$
          (It will be noted that there is an incomplete chain in the network, made up of the single arc $(1,3)$).
step 1   Label node 1 $(-, \infty)$.
step 2   Arc $(1,3)$ is case (a).
step 3   Label node 3 $(1,\min(\infty,3-0)) = (1,3)$.
step 2   Arc $(2,3)$ is case (b).
step 3   Label node 2 $(-3,\min(3,5)) = (-3,3)$.
step 2   Arc $(3,4)$ is case (a).
step 3   Label node 4 $(3,\min(3,6-5)) = (3,1)$, node 4 is now labelled.

step 4   Increase flow in the chain $(1,3)$, $(3,4)$ by 1
          so $x(3,4) = 5 + 1 = 6$
            $x(1,3) = 0 + 1 = 1$
          (again there is an incomplete chain $(1,3)$, $(2,3)$).
step 1   Label node 1 $(-, \infty)$.
step 2   Arc $(1,3)$ is case (a).
step 3   Label node 3 $(1,\min(\infty,3-1)) = (1,2)$.
step 2   Arc $(2,3)$ is case (b).
step 3   Label node 2 $(1,\min(\infty, 8-5)) = (1,3)$.
step 2   Arc $(2,4)$ is case (a).
step 3   Label node 4 $(2,\min(2,7-3)) = (2,2)$, node 4 is now labelled.
step 4   Increase flow in the chain $(1,3)$, $(2,3)$, $(2,4)$ by 2
          so $x(2,4) = 3 + 2 = 5$
            $x(2,3) = 5 - 2 = 3$
            $x(1,3) = 1 + 2 = 3$.
step 1   Label node 1 $(-, \infty)$.
step 2   There are no arcs of case (a) or case (b).
step 5   Stop with the maximal flow from 1 to 4,
          this maximal flow is $x(1,2) + x(1,3) = 8 + 3 = 11$ units.

4.2.2 Notes

Cut-sets in the network
In considering the capacity of the network, it is very useful in define the idea of a cut-set, and its capacity. A **cut-set** is a set of arcs, which connect a set X of nodes of N to its complement. The source is in the set X while the sink is not. So, using $\bar{X}$ to denote the complement of X, the cut-set created by set X is the set of arcs

$$(X,\bar{X}) = \{(i,j) \ : \ i \in X, \ j \in \bar{X}\} \subset A \ .$$

The capacity of this cut-set is the sum of the capacities of the arcs in it.

As flow has to pass from s to t, via various paths, each unit of flow must cross from X to $\bar{X}$. Thus the capacity of the cut-set must be least as large as the maximal flow from s to t through the network, and this is true for all cut-sets. Moreover, the value of the maximal flow is equal to the minimum capacity of all cut-sets in the network. This result is known as the **max-flow-min-cut theorem.**

Proof of the max-flow-min-cut theorem
Let f be the maximum flow found by the algorithm, and let S be the set of labelled nodes when step 5 is reached. $(S,\bar{S})$ defines a cut-set in the network, since s is in S and t is not (otherwise there would be a flow-augmenting chain).

Consider arcs in $(S, \bar{S})$ and in $(\bar{S}, S)$. The net sum of all the flows between S and $\bar{S}$ must be f, that is,

$$\sum_{\substack{i \in S \\ j \in \bar{S}}} x(i,j) - \sum_{\substack{k \in \bar{S} \\ l \in S}} x(k,l) = f$$

For each arc (i,j) in $(S, \bar{S})$, $x(i,j) = u(i,j)$, or else node j could be labelled with flow from node i. Therefore:

$$\sum_{\substack{i \in S \\ j \in \bar{S}}} x(i,j) = \sum_{\substack{i \in S \\ j \in \bar{S}}} u(i,j) = \text{capacity of } (S, \bar{S})$$

Similarly, for each arc (k, l) in $(\bar{S}, S)$, $x(k,l) = 0$, or else node k could be labelled as a reverse arc from node l. So:

$$\sum_{\substack{k \in \bar{S}, \ l \in S}} x(k,l) = 0$$

Bringing these two together:

$$\sum_{\substack{i \in S, \ j \in \bar{S}}} x(i,j) - \sum_{\substack{k \in \bar{S}, \ l \in S}} x(k,l) = \text{capacity of } (S, \bar{S}) - 0 = f$$

And thus the maximum flow equals the capacity of this cut-set, as the maximum flow is less than or equal to the capacity of all cut-sets, the capacity of the cut-set $(S, \bar{S})$ is minimal, and the result is achieved.

The cut-set $(S, \bar{S})$ need not be unique.

Several sources and sinks
The maximal flow algorithm can be readily adapted to other problems which are related in the sense of seeking for particular flow patterns in a network with upper bounds on the arcs. One such problem is that of finding the maximum flow through a network, where there are several sources and/or sinks. It would be possible to solve this problem by finding the maximal flow from one source to one sink, and then reducing the arc capacities by the flows in them, finding the optimal flow between another pair of nodes and so on until all pairs have been investigated. But this is not really the most straightforward way of tackling the problem. It is better to extend the network by the addition of two new nodes, which act as a 'super-source' and a 'super-sink'. The super-source will be linked to all the sources by means of arcs with very large upper bounds, and similar arcs will link the sinks to the super-sink. Then the maximal flow from super-source to super-sink will yield the solution of the original problem. The flows in the newly created arcs are the corresponding flows from each source or to each sink.

There are several possible adaptations to the network which may be appropriate for particular optimisation problems. It may be desirable in such a problem, to balance the arcs so that all are supplying equal flows to the network. In such a case, an iterative approach can be used. Instead of setting the capacity of the arcs from the super-source to be very large, the problem will initially be solved with a very small capacity on each one. If there is a solution to the maximal flow problem for which these arcs are all being used fully, then the capacity of each can be increased; this continues until there is no solution using the arcs to their full capacity. The solution found before this will correspond to the problem of achieving balance.

Upper bounds on flows through nodes

In some flow problems, such as water distribution problems, there is a limit to the flow which can be routed through certain nodes. To solve problems where there are such node-capacity limits, it is necessary to modify the network. This is done by extending the limited nodes to be in two parts, one node for the flow entering the original node, and one node for flow leaving the original node. These two can be linked by an arc whose capacity is the capacity of the node, and the maximal flow through the modified network found.

The most critical arc in a network

It is sometimes desirable, given the solution to a maximal flow problem, to speculate as to the effect of the removal of one of the arcs. In particular, for the controller of the network being modelled, the most important arc might be that whose removal would reduce the maximal flow by the greatest amount. To find this, a two-stage process is necessary. First, the arcs are ranked according to the flow which they carry, and those arcs which appear in minimal cut-sets are identified. If the arc with the largest flow is also in such a cut-set, then this arc is the most important in the network. But if not, the arcs with flow larger than the flow in the largest member of a minimal cut-set must be deleted one-at-a-time from the network, and the effects of each deletion recorded. The arc with the greatest effect can thus be found. Examples of each situation are shown below. (Figs. 4.7 and 4.8.)

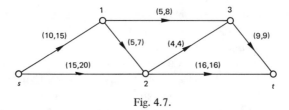

Fig. 4.7.

In Fig. 4.7 the minimal cut-set is defined by $S = \{s,1,2,3\}$. The arc with the largest flow is $(2,t)$ with capacity 16 and this is in the cut-set.

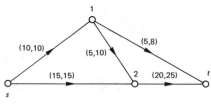

Fig. 4.8.

In Fig. 4.8 the minimal cut-set is defined by $S = \{s\}$. The arc with the largest flow is $(2,t)$ which is not in the cut-set. But if it is deleted, the flow will fall from 25 units to 8 units, which is more than would happen if arc $(s,2)$ were to be deleted.

4.2.3 Implementation

The formal statement of the algorithm presents some minor problems when it has to be transformed into a computer program. The labelling procedure assumes that labels can be created as it were 'out of thin air'. Rather than attempt actually to do this, it is better to use a method which assigns labels to nodes so that in the cases where there is no label, this may be readily detected. This could be done, as in some of the shortest path algorithms, using logical flags for each node, but it is as effective to give each node a label which is wholly zero until it has values assigned to it within the algorithm.

As usual, the network is readily stored in a series of one-dimensional arrays; each arc may be represented by its start and finish node, its capacity, and currently assigned flow. This last is varied by the program. The labels on nodes may also be stored in two arrays, corresponding to the two parts of the label (although an array with one row for each node and two columns could also be used).

With this data structure, the two programs can follow the algorithm fairly closely, with the use of GOTO statements following logical tests in BASIC replaced by the more flexible testing mechanisms of PASCAL.

There are other problems associated with this straightforward translation from algorithm to program. In step 2 of the method, there are three distinct types of arc to be identified, namely: forward arcs, reverse arcs and everything else. This third group is identified tacitly in the formal statement given above, and is made up of arcs linking two unlabelled nodes and arcs linking two labelled nodes, as well as forward arcs with no spare capacity and reverse arcs with zero flow. The restricted use of logical tests in BASIC means that it is easier to identify some of the members of this group first, and then to find members of the first and second groups from what remains. To do this, extensive use is made of the numerical values of the second part of the labels on the ends of the arc, $B(E)$ and $B(F)$. For any arc which has both ends labelled, both $B(E)$

and $B(F)$ will be non-zero, and so their product will be non-zero as well; for arcs between a labelled and an unlabelled node, and for arcs between two unlabelled nodes, this product will be zero. By testing whether the product is zero or not, some members of the third group can be eliminated. Considering the arcs for which the product $B(E) * B(F) = 0$, the only arcs for which the sum is also zero will be members of the third group as well, so these may be eliminated from further consideration. The arcs which remain will have either $B(E) = 0$ or $B(F) = 0$. Forward arcs can be found by examining the product $B(E) * (U(C) - X(C))$. This will be non-zero only when node E has been labelled. and there is some spare capacity in the arc C. In the same way, reverse arcs can be identified as those for which $B(F) * X(C)$ is non-zero.

Step 4 also presents a problem, since it is necessary to use the labels to find the end nodes of an arc in order to change its flow. This involves finding the orientation of the arc, and its position in the list of arcs. Rather than have to scan the list of arcs, it is more efficient to amend the formal algorithm so that the first part of the label on a given node is the number of the arc which leads to it in the chain, set to be positive for forward arcs, and negative for reverse arcs.

The PASCAL program unites steps 2 and 3, using tests on the arcs which mirror exactly the formal statements of the algorithm, rather than the circuitous tests of the BASIC version; but in most other respects, the two codings are identical.

```
program prog4p1(input,output);
  const maxnodes = 30;
        maxarcs  = 50;
        inf = 9999;
  type  ar = array[1..maxarcs] of integer;
 nodear = array[1..maxnodes] of integer;
  var   i,j,u,x : ar;
        a,b : nodear;
        n,arcs,s,t,t1,l,flo : integer;
        newlabel : boolean;

function min(a,b : integer): integer;
{min is the smaller of a and b}
begin
    if (a<b) then min := a else min := b
end;


procedure readnet;
{ reads the network, with capacities }
var t8 : integer;
begin
repeat
    write('How many nodes, how many arcs?');
    readln(n,arcs);
    if (n>maxnodes) then
      writeln('too many nodes:   maxnodes
##    should be changed');
```

```
      if (n<2) then
        writeln('number of nodes should be more
##      than 1:  try again');
      if (arcs>maxarcs) then
        writeln('too may arcs: maxarcs should
##      be changed');
      if (arcs<1) then
        writeln('number of arcs should be
##      positive:  try again');
      if (n>arcs+1) then
        writeln('there should be at least',
##      n-1:4,' arcs: try again')
  until ((n in [1..maxnodes])
and(arcs in [1..maxarcs])and(n<=arcs));

  writeln('Enter arcs with start node finish
##      node capacity');

  for t8 := 1 to arcs do
  repeat
    write('Arc number',t8:3,'  ');
    readln(i[t8],j[t8],u[t8]);
    if ((i[t8]>n)or(j[t8]>n)) then
  writeln('Node number too high:  try again');
    if ((i[t8]<1)or(j[t8]<1)) then
  writeln('Node number too low:  try again');
    if (u[t8]<1) then
      writeln('capacity should be positive:
##      try again');
    if (i[t8]=j[t8]) then
      writeln('Start and finish must be
##      different:  try again')
  until (((i[t8]in[1..n])and(j[t8]in[1..n])
        and(i[t8]<>j[t8])and(u[t8]>0));
  end {readnet};
  begin
  {Maximum flow algorithm
  variables used:
  i : vector of start nodes for arcs
  j : vector of finish nodes for arcs
  u : upper bound on flows in arcs
  x : currently assigned flows in arcs
  a : vector of labels for nodes,
##      corresponding to the arc
          number which may be used to send flow
##      from the preceding
  node in the flow augmenting chain from s
  b : vector of labels for nodes,
##      corresponding to the maximum
          flow which can be sent along the flow
##      augmenting chain
  n : number of nodes
  arcs : number of arcs
  s : source node
  t : sink node
  l : variable used for identifying arcs in
##      the flow augmenting
          chan from s to t
  flo : amount of flow from s to t
  t1 : loop counter
  newlabel : true when a pass through the
##      arcs finds a node which
                  may be labelled
  }
  writeln('Maximum flow algorithm');
  readnet;
```

```
   repeat
     write('Enter start node and finish node
##     for flow');
     readln(s,t);
     if (s=t) then
       writeln('Start and finish should be
##   different:  try again');
     if ((s>n)or(t>n)) then
writeln('Node number too high:  try again');
     if ((s<1)or(t<1)) then
writeln('Node number too low:  try again')
   until ((s<>t)and(s in [1..n])and(t in
##     [1..n]));

   for t1 := 1 to arcs do x[t1] := 0;
   flo := 0;
   {                ******
                    STEP 1
                    ******
   }
   repeat
     for t1 := 1 to n do
     begin
       a[t1] := 0;
       b[t1] := 0
     end;
     b[s] := inf;
   {                ******
                    STEP 2
                    ******
   }
     repeat
       t1 := 1;
       while ((b[t]=0)and(t1<=arcs))do
       begin
         newlabel := false;
         if ((b[i[t1]]>0)and(b[j[t1]]=0)
             and(x[t1]<u[t1]))
         then
         begin
           newlabel := true;
           a[j[t1]] := t1;
b[j[t1]] := min(b[i[t1]],u[t1]-x[t1])
         end;
         if ((b[i[t1]]=0)and(b[j[t1]]>0)
             and(x[t1]>0)) then
         begin
           newlabel := true;
           a[i[t1]] := -t1;
           b[i[t1]] := min(b[j[t1]],x[t1])
         end;
if newlabel then t1 := 1 else t1 := t1+1
       end
     until ((b[t]>0)or(not newlabel));

     if (b[t]>0) then
     begin
   {                ******
                    STEP 4
                    ******
   }
       l := a[t];
       while (i[abs(l)]<>s) do
       begin
         if (l>0) then
         begin
           x[l] := x[l] + b[t];
```

```
          l := a[i[abs(l)]]
      end
      else
      begin
        x[-l] := x[-l] - b[t];
        l := a[j[abs(l)]]
      end
    end;
    if (l>0) then x[l] := x[l] + b[t]
             else x[-l] := x[-l] - b[t];
    flo := flo + b[t]
  end
until (not newlabel);
writeln('Maximum flow found');
writeln('The flow is ',flo,' and is
##   distributed in arcs as follows:');
writeln('start node finish node capacity
##     flow');
for t1 := 1 to arcs do
writeln(i[t1]:10,j[t1]:10,u[t1]:10,x[t1]:10);
writeln;
writeln('the nodes in the cut set are:');
for t1 := 1 to n do
  if (b[t1]>0) then write(t1:4);
writeln;
end.
```

```
90 REM PROG4P2
100 DIM I(100),J(100),U(100),X(100),
    A(100),B(100)
110 Z=100
120 PRINT "MAXIMUM FLOW ALGORITHM"
130 REM              VARIABLES USED
140 REM            **************
150 REM M - NUMBER OF NODES IN NETWORK
160 REM N - NUMBER OF ARCS
170 REM Z - MAXIMUM ALLOWED VALUES FOR
    M AND N:  CAN BE ALTERED
180 REM S - NUMBER OF NODE ACTING AS
    SOURCE OF FLOW
190 REM T - NUMBER OF NODE ACTING AS SINK
200 REM V - TOTAL FLOW THROUGH NETWORK
210 REM G - EXTRA FLOW SENT ALONG
    FLOW-AUGMENTING PATH
220 REM I(C) - NODE NUMBER FOR START OF
    ARC C
230 REM J(C) - NODE NUMBER FOR FINISH
    OF ARC C
240 REM U(C) - CAPACITY (UPPER BOUND)
    OF ARC C
250 REM X(C) - FLOW ALONG ARC C
260 REM A(D) - FIRST PART (PREDECESSOR
    NODE) OF LABEL ON NODE D
270 REM B(D) - SECOND PART (POTENTIAL
    FLOW) OF LABEL ON NODE D
280 REM C,D,E,F,H - WORK SPACE, OR LOOP
    COUNTERS
290 REM        THE VERSION OF BASIC USED
    FOR THIS PROGRAM REQUIRED
300 REM        THAT FOUR PARAMETERS BE
    USED IN THE FUNCTION "MIN"
310 REM        IF THIS IS NOT THE CASE
    IN THE VERSION BEING USED,
```

```
320 REM          THEN AN APPROPRIATE
    CHANGE SHOULD BE MADE.
330 PRINT "ENTER NUMBER OF NODES,
    NUMBER OF ARCS   ";
340 INPUT M,N
350 IF M>N THEN 1390
360 IF M>Z THEN 1410
370 IF N>Z THEN 1440
380 IF N<1 THEN 1460
390 IF M<1 THEN 1480
400 PRINT "ENTER SOURCE NODE, SINK NODE   ";
410 INPUT S,T
420 IF S=T THEN 1500
430 IF S>M THEN 1520
440 IF T>M THEN 1540
450 IF S<1 THEN 1560
460 IF T<1 THEN 1580
470 V=0
480 PRINT "INPUT DETAILS FOR EACH ARC:"
490 PRINT "IN THE ORDER: START NODE,
    FINISH NODE, CAPACITY"
500 FOR C = 1 TO N
510 PRINT "ARC NUMBER ";C;
520 INPUT I(C),J(C),U(C)
530 IF I(C)>M THEN 590
540 IF I(C)<0 THEN 610
550 IF J(C)>M THEN 590
560 IF J(C)<0 THEN 610
570 IF U(C)<0 THEN 630
580 GOTO 650
590 PRINT "NODE NUMBER MUST BE LESS THAN";M
600 GOTO 520
610 PRINT "NODE NUMBER MUST BE POSITIVE"
620 GOTO 520
630 PRINT "CAPACITY MUST BE GREATER
    THAN OR EQUAL TO ZERO"
640 GOTO 520
650 X(C)=0
660 NEXT C
670 REM              **********
680 REM              * STEP 1 *
690 REM              **********
700 FOR D = 1 TO M
710 A(D)=0
720 B(D)=0
730 NEXT D
740 B(S)=9999
750 REM              **********
760 REM              * STEP 2 *
770 REM              **********
780 FOR C = 1 TO N
790 E=I(C)
800 F=J(C)
810 IF B(E)*B(F) <> 0 THEN 850
820 IF B(E)+B(F)=0 THEN 850
830 IF B(E)*(U(C)-X(C)) > 0 THEN 900
840 IF X(C)*B(F)> 0 THEN 940
850 NEXT C
860 GOTO 1220
870 REM              **********
880 REM              * STEP 3 *
890 REM              **********
900 A(F)=E
910 H=U(C)-X(C)
920 B(F)=MIN(H,H,H,B(E))
930 GOTO 960
```

```
940 A(E)=-F
950 B(E)=MIN(B(F),B(F),B(F),X(C))
960 IF B(T)=0 THEN 780
970 REM                 **********
980 REM                 * STEP 4 *
990 REM                 **********
1000 G=B(T)
1010 V = V+G
1020 F=T
1030 E=ABS(A(F))
1040 FOR C = 1 TO N
1050 IF A(F)>0 THEN 1090
1060 IF I(C)<>F THEN 1140
1070 IF J(C)<>E THEN 1140
1080 GOTO 1110
1090 IF I(C)<>E THEN 1140
1100 IF J(C) <>F THEN 1140
1110 IF X(C)+SGN(A(F))*G<0 THEN 1140
1120 IF X(C)+SGN(A(F))*G>U(C) THEN 1140
1130 GOTO 1150
1140 NEXT C
1150 X(C)=X(C)+SGN(A(F))*G
1160 F=E
1170 IF F<> S THEN 1030
1180 GOTO 700
1190 REM                 **********
1200 REM                 * STEP 5 *
1210 REM                 **********
1220 PRINT "TOTAL FLOW IS ":V
1230 PRINT "AN OPTIMAL FLOW PATTERN IS:"
1240 PRINT "START NODE     FINAL NODE
     CAPACITY        FLOW"
1250 FOR C=1 TO N
1260 PRINT I(C),J(C),U(C),X(C)
1270 NEXT C
1280 PRINT
1290 PRINT "THE LABELS ON THE NODES ARE
     AS FOLLOWS:"
1300 PRINT "NODE NUMBER      PREDECESSOR
     POTENTIAL FLOW"
1310 FOR D = 1 TO M
1320 PRINT D,A(D),B(D)
1330 NEXT D
1340 PRINT "NODES WITH NON-ZERO
     POTENTIAL FLOW ARE IN THE CUT SET"
1350 STOP
1360 REM
1370 REM    THE MESSAGES THAT FOLLOW
     CORRESPOND TO ERRORS IN INPUT
1380 REM
1390 PRINT "NUMBER OF NODES MUST BE
     LESS THAN NUMBER OF ARCS"
1400 GOTO 330
1410 PRINT "NUMBER OF NODES TOO LARGE:
     ALTER DIMENSIONS AND Z"
1420 PRINT "IN LINES 100 AND 110"
1430 STOP
1440 PRINT "NUMBER OF ARCS TOO LARGE:
     ALTER DIMENSIONS AND Z"
1450 GOTO 1420
1460 PRINT "THERE MUST BE AT LEAST ONE ARC"
1470 GOTO 330
1480 PRINT "THERE MUST BE AT LEAST ONE NODE"
1490 GOTO 330
1500 PRINT "SOURCE AND SINK MUST BE
     DIFFERENT NODES"
```

```
1510 GOTO 400
1520 PRINT "SOURCE NUMBER MUST BE LESS
     THAN";M
1530 GOTO 400
1540 PRINT "SINK NUMBER MUST BE LESS THAN";M
1550 GOTO 400
1560 PRINT "SOURCE NUMBER MUST BE POSITIVE"
1570 GOTO 400
1580 PRINT "SINK NUMBER MUST BE POSITIVE"
1590 GOTO 400
```

4.3 THE PATH OF MAXIMUM FLOW

In certain circumstances, instead of wishing to find the maximum flow which can be sent from source s to sink t by any combination of arcs, it is desirable to find the path whose capacity is largest. Then, flow can be sent from s to t along this path, and this flow need not be divided into parts. This problem resembles the maximum flow problem in many ways, but it may be more readily solved using the shortest path algorithm.

In the chapter on shortest paths, it was noted that the labelling algorithm could be adapted to any problems which involved the comparision of two or more paths for some numerical property, followed by the selection of one or the other of these for retention. The capacity of the path is such a property. In essence, the method for such problems follows the Dijkstra algorithm. Where this assigns a zero label to the start node, to represent distance, an infinite label is given to represent unlimited capacity. Where a total distance is compared, and the smaller selected, two capacities are compared and the larger taken. Labels are assigned successively, and eventually the sink will have a permanent label.

4.4 THE OUT-OF-KILTER ALGORITHM

There are several situations in which the objective is not to find a flow which satisfies one criterion only. It may be that flows in arcs are constrained by upper and lower bounds, which must be satisfied. There may be a cost-per-unit of flow. A general problem is that of finding a flow in a network which has minimal total cost, subject to upper and lower limits on the flows in individual arcs. This problem is known as the **minimal-cost-flow problem.** The maximal flow algorithm is a special case of this general problem; so in another sense, is the shortest route problem, since this may be regarded as a problem of sending one unit of flow (the traveller) from one city to another at minimal cost (distance).

Generally, a further condition is added to the constraints which have been given on each arc; this is that flow is conserved at every node of the network, or that there is no exogeneous flow in the system. This condition was present for the maximal flow problem, but there it was only applied to nodes other than the source and the sink. In the general minimal-cost-flow problem, there are no sources and sinks. This is not such a serious limitation on the applicability of this problem as may at first appear. For instance, in the maximal flow problem, a network with no exogeneous flow can be created by introducing an arc from sink to source, so that the flow circulates around the network.

In formal consideration of the problem, there is a given network whose graph is (N,A); associated with each of the arcs in A, there is a cost per unit of flow $c(i,j)$ in arc (i,j), and the flow in that arc is $x(i,j)$ which satisfies $0 \leqslant l(i,j) \leqslant x(i,j) \leqslant u(i,j)$. The objective is to find a set of flows which minimises $\sum\limits_{\text{all arcs}} x(i,j) * c(i,j)$. The cost per unit of flow is any integer, and the bounds on flow are any non-negative integers.

The approach to be taken in solving this problems derives many of its ideas from linear programming, although knowledge of this is not necessary for the method to be used successfully. Given a linear program whose constraints are all equalities, then the solution to the linear programming problem is not changed if multiples of any or all of these constraints are added to the objective function, to give a new objective function and the same constraints. Adding such multiples will, of course, change the coefficients of the objective function, and will, if suitably chosen, make for a simpler one. (In effect, simplex algorithms for linear programs do select such multiples.)

The minimal-cost-flow problem is a linear-programming problem. It has an objective function which is linear, and all the constraints, both on flow size and on flow conservation, are linear equality constraints, when slack variables have been introduced. This problem has a very simple form, as the constraints are of three types only, corresponding to:

(1) flow conservation at nodes
(2) lower bounds on flows
(3) upper bounds on flows

At node j, the flow conservation equation has the form:

$$\sum_i x(i,j) = \sum_k x(j,k), \text{ or } -\sum_i x(i,j) + \sum_k x(j,k) = 0$$

In the node-arc incidence matrix the row corresponding to node j has zero entries except for those arcs which terminate at j, and the entries are +1 for arcs whch have their start node at j, (that is, arcs (j,k)) and entries of -1 for those whose end node is j. So, the equality constraint above corresponds to the product of the jth row of the node-arc incidence matrix with the vector of flows in the network. Since this is true for all the nodes in the network, this set of constraints may be regarded as the product of the node-arc incidence matrix with the vector of flows having as its value a zero-vector.

In arc (i,j), the lower bound constraint has the form: $x(i,j) \geqslant l(i,j)$, which becomes, after the introduction of a slack variable $s(i,j)$: $x(i,j) - s(i,j) = l(i,j)$. In the same way, the upper bound constraint can be written as: $x(i,j) + t(i,j) = u(i,j)$, with the introduction of a slack variable $t(i,j)$.

Bringing all these three types together, the matrix for the linear programming problem may be drawn up, in the form shown in Fig. 4.9. Here, the blocks which are wholly zero are shown as 0, and the square matrices whose diagonal entries are all +1, with zero everywhere else are shown as 1. The size of the problem will be: number of rows: + number of nodes + 2 * number of arcs; number of columns: 3 * number of arcs.

x	s	t	
E	0	0	0
1	−1	0	l
1	0	+1	u

C	0	0

Fig. 4.9 – The matrix for the minimal-cost-flow problem (E is the node-arc incidence matrix).

Using the idea introduced earlier, that the objective function may be modified by the addition of suitable multiples of the constraints, three types of multipliers will be introduced, corresponding to the three types of constraint. Corresponding to the flow conservation at node j, a multiplier π_j will be used. Corresponding to the lower bound on (i,j) will be a multiplier $\lambda(i,j)$, and to the upper bound a multiplier $\eta(i,j)$. In the objective function, there will be three terms corresponding to the arc. These will change from:

$$c(i,j)*x(i,j) + 0*s(i,j) + 0*t(i,j)$$

to: $\quad \{c(i,j) + \pi_i - \pi_j + \lambda(i,j) + \eta(i,j)\} *x(i,j) - \lambda(i,j)*s(i,j) + \eta(i,j)*t(i,j)$.

Now suppose the flows x form an optimal solution. Then linear programming theory yields that one condition for this is that:

(a) $\quad c(i,j) + \pi_i - \pi_j + \lambda(i,j) + \eta(i,j) = 0$

and (b) the coefficient of $s(i,j)$ (or $t(i,j)$) is zero when $s(i,j)$ (or $t(i,j)$) is positive, and is non-negative when $s(i,j)$ (or $t(i,j)$) is zero.

These two conditions ensure that the coefficients of basic variables are zero, and of non-basic variables are non-negative. ($x(i,j)$ is always assumed to be basic.)

Condition (b) can be rewritten in terms of $x(i,j)$ and its bounds as:

(b*) $\lambda(i,j) = 0$ for $x(i,j) > l(i,j)$
$\quad\,\,\, \lambda(i,j) \leqslant 0$ for $x(i,j) = l(i,j)$

$\quad\,\,\, \eta(i,j) = 0$ for $x(i,j) < u(i,j)$
$\quad\,\,\, \eta(i,j) \geqslant 0$ for $x(i,j) = u(i,j)$

Using these conditions, it is possible to identify three possible ways in which they may be satisfied. Each of these gives rise to a condition on the value of the sum $c(i,j) + \pi_i - \pi_j$ and the corresponding value of $x(i,j)$.

   (i)   If $x(i,j) = l(i,j)$ then $\lambda(i,j) \leqslant 0$ and $\eta(i,j) = 0$. By choosing $c(i,j) + \pi_i - \pi_j \geqslant 0$, and setting $\lambda(i,j) = -(c(i,j) + \pi_i - \pi_j)$ condition (a) will be satisfied.
   (ii) If $l(i,j) < x(i,j) < u(i,j)$ then $\lambda(i,j) = 0$ and $\eta(i,j) = 0$. Condition (a) will be satisfied if $c(i,j) + \pi_i - \pi_j = 0$.
   (iii) If $x(i,j) = u(i,j)$ then $\eta(i,j) \geqslant 0$ and $\lambda(i,j) = 0$. By choosing $c(i,j) + \pi_i - \pi_j \leqslant 0$ and setting $\eta(i,j) = -(c(i,j) + \pi_i - \pi_j)$ condition (a) will be satisfied.

(In the case $l(i,j) = u(i,j)$, then the only feasible flow will be $x(i,j) = l(i,j) = u(i,j)$, and so $s(i,j)$ and $t(i,j)$ will both be zero. $\lambda(i,j)$ and $\eta(i,j)$ can be chosen arbitrarily, and ($c(i,j) + \pi_i - \pi_j$) can take any value at all).

The term $c(i,j) + \pi_i - \pi_j$ may be thought of as a single variable; when the three cases considered above are plotted on a graph with this variable as the vertical axis, plotted against the flow as the horizontal axis, then it is evident that a flow in arc (i,j) corresponds to a point on the bold line shown when an optimal flow pattern has been established. This line is referred to as the 'kilterline' for the arc (i,j), and if an arc has a flow, and has node multipliers π_i and π_j which correspond to a position on the line, then it is deemed to be 'in kilter'. If it falls in some other position in the diagram, then it is 'out-of-kilter'. (Fig. 4.10).

It is a necessary condition for the set of flows to be optimal that each arc is in kilter. This amounts to systematically selecting the flows and the multipliers for the nodes (that is, the node conservation equations) so that the point ($x(i,j)$, $c(i,j) + \pi_i - \pi_j$) is on the kilter line. In order that this may be achieved, an algorithm due to Ford and Fulkerson [19] can be used. In this approach, there are two parts; one makes changes to the flows in selected arcs, which has the effect of moving an arc's position in its kilter diagram horizontally, while the other makes changes to the multipliers, for selected nodes, which has the effect of moving the position of some of the arcs up or down in the kilter diagram. These two parts are separate, so that it is not possible for a change to be made which simultaneously moves an arc's position horizontally and vertically.

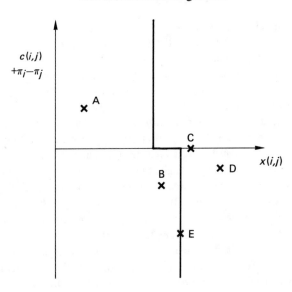

Fig. 4.10 – A, B, C, D are out-of-kilter. E is in kilter.

Since it is necessary for the solution that all the arcs in the network are in kilter, the algorithm first seeks for the arcs which are out-of-kilter. Such arcs can possibly be made in kilter by a suitable change in the flow which they carry, and so the first part of the out-of-kilter algorithm seeks to find a flow-augmenting chain which will help to bring the selected arc into kilter, or, if this is not possible, at least closer to the kilter line. Where there is no such flow-augmenting chain, then the second part of the algorithm is invoked, and a change determined to the π-multipliers for selected nodes. This has the effect either of bringing some arcs into kilter, or of enabling them to be used in a flow-augmenting path. A labelling algorithm which is very similar to that used in the maximum flow problem is used, and the set of nodes which cannot be labelled at any given time (because of their flows and/or their values of $c(i,j) + \pi_i - \pi_j$) is the set whose π-multipliers are changed.

The algorithm requires a set of initial values for the flows and for the π-multipliers. These need not be feasible; it is common, and often convenient, to use zero flow in all arcs as the initial condition, and to set all the π-multipliers to zero as well. However, this is not necessary. When the minimal-cost-flow problem being studied has been slightly changed from some other one, then the optimal flows and multipliers from the previous problem can be used as the starting values for the new problem. The only requirement on the flows is that the flow-conservation equations be satisfied. If there is an inbalance at any node, then this cannot be corrected within the algorithm; generally this is not desirable.

Following the initialisation of flows, the main step of the algorithm is entered. This searches for an out-of-kilter arc, and then seeks to bring it into kilter; if this cannot be done, then there is no feasible solution to the problem, and so the algorithm terminates.

As has been described above, there are two ways of changing the position of an arc in its kilter diagram. The first to be used is a labelling algorithm to change the flow in the arc which is out of kilter.

Suppose the arc (s,t) is out of kilter. It may be out of kilter because its flow is too low for its current value of $c(s,t) + \pi_s - \pi_t$; or its flow may be too high. These two cases may be illustrated diagrammatically, as in Figs. 4.11 and 4.12. If the flow is not large enough, then the arc may be brought into kilter by increasing the flow from s to t; since flow must be conserved, a flow-augmenting chain from t to s must be found. If the flow is too large, then the arc may be brought into kilter by reducing the flow from s to t, and this requires a chain from s to t to contain the diverted flow through the network.

Thus, in either case, a flow-augmenting chain is needed, and it is conventional to reverse the labels for arcs in the first case, so that the flow-augmenting chain is always from s to t. The conditions for including arcs in flow-augmenting chains are very similar to those encountered for the maximum flow problem, except that the arc's position in its kilter diagram must be used to determine whether it may be used in the chain, and if so, what the extra flow which can be directed along it might be. Underlying the rules for the inclusion of arcs is one dominant proviso. No arc which is in kilter should ever be taken out-of-kilter. A subsidiary rule is that arcs should be moved towards their kilter line, and not away from it.

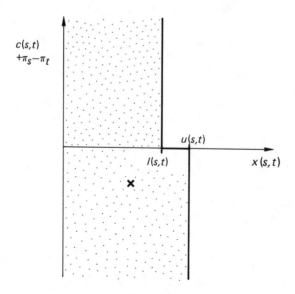

Fig. 4.11.

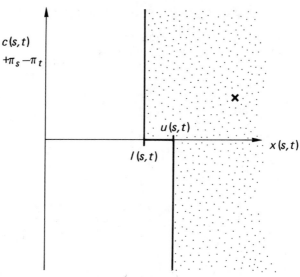

Fig. 4.12.

The chain may include both forward arcs and reverse arcs. A forward arc (i,j) may be used in the chain if:

either (a) $c(i,j) + \pi_i - \pi_j \leqslant 0$ and $x(i,j) \leqslant u(i,j) - 1$

or   (b) $c(i,j) + \pi_i - \pi_j > 0$ and $x(i,j) \leqslant l(i,j) - 1$

(*n.b.* $x(i,j) \leqslant u(i,j) - 1$ corresponds to the inequality $x(i,j) < u(i,j)$ with both $x(i,j)$ and $u(i,j)$ integral, and similarly for $x(i,j) \leqslant l(i,j) - 1$).

The kilter positions for forward arcs are as shown in Fig. 4.13.

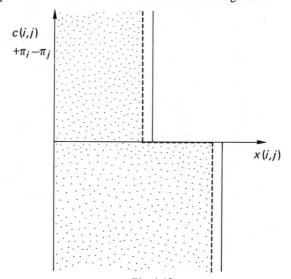

Fig. 4.13.

Similarly, a reverse arc (i,j) may be used in the chain if:

either (a) $c(i,j) + \pi_i - \pi_j \geqslant 0$ and $x(i,j) \geqslant l(i,j) + 1$
or   (b) $c(i,j) + \pi_i - \pi_j < 0$ and $x(i,j) \geqslant u(i,j) + 1$.
The kilter positions for such arcs are as shown in Fig. 4.14.

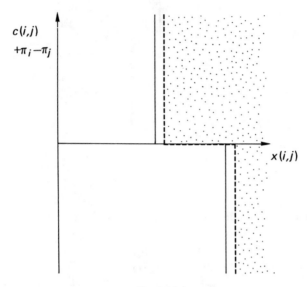

Fig. 4.14.

The arcs which are used in flow-augmenting chains are generally themselves out-of-kilter, except for those arcs for which $c(i,j) + \pi_i - \pi_j = 0$.

The labelling algorithm which is used to find the chain is essentially the same as that which is used in the maximal flow algorithm, with a two part label to define the preceding node of the chain and the amount of extra flow which can be sent along the chain to the node being labelled. However, the mechanism for calculating the amount of extra flow is different, as it involves tests on the position of arcs in their kilter diagram. Since an arc may be moved up to, but not beyond its kilter line, the amount of extra flow which can be sent along forward arcs is not always $u(i,j) - x(i,j)$ for a forward arc (i,j). In those cases where $c(i,j) + \pi_i - \pi_j > 0$, the extra flow will only be $l(i,j) - x(i,j)$, since any extra flow change beyond this would move the arc beyond the kilter line. So, the labelling rules limit the changes of flow in any arc, and hence in the whole chain, by forbidding changes of too great a magnitude.

In order that a flow-augmenting chain may be found, the node s is given a label according to the same rules (the change in flow in the chain must not take the arc between s and t out of kilter beyond its kilter line) and the chain developed. When node t has been labelled, then the flow in the chain is increased by the amount b_t. The change in flow is such that no arc is taken out-of-kilter, and all arcs which were out-of-kilter are brought nearer to their kilter lines. Once the flow change has been completed, then all the labels are discarded, and the arc between s and t is tested once more.

It may happen that it is not possible to find a flow-augmenting chain to label node t. In this case, the second method of changing the position of arcs in their kilter diagram has to be used. This changes the node multipliers on all unlabelled nodes, by adding a quantity δ to each one. The effect of this is only seen on those arcs which link labelled and unlabelled nodes. For arcs linking two labelled nodes, then the π-multipliers are not changed, so that the expression $c(i,j) + \pi_i - \pi_j$ is not altered. For arcs linking two unlabelled nodes, the value of the expression is also unaltered, since the δ which is added to π_i is balanced by the δ which is added to π_j. The value of δ is chosen so that at least one of the arcs linking labelled and unlabelled nodes is moved onto the horizontal portion of the kilter line, and no arc is moved beyond this. So δ is defined as the minimum of $(\ |\ c(i,j) + \pi_i - \pi_j\ |\ )$ taken over all arcs (i,j) which satisfy either:

(a) i labelled, j not labelled and $l(i,j) \leqslant x(i,j) < u(i,j)$ (note the form of the latter inequality).

(b) j labelled, i not labelled and $l(i,j) < x(i,j) \leqslant u(i,j)$ (note this form also)

(In case (a), $c(i,j) + \pi_i - \pi_j > 0$, or otherwise j could be labelled in the chain with flow from i; in case (b), $c(i,j) + \pi_i - \pi_j < 0$, or else i could be labelled from j). These correspond to the two situations shown in Figs. 4.15 and 4.16.

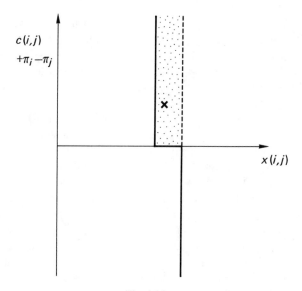

$c(i,j)$

$+\pi_i - \pi_j$

$x(i,j)$

Fig. 4.15.

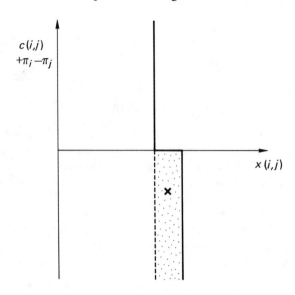

Fig. 4.16.

In case (a), adding δ to π_j moves the position of the arc in the kilter diagram towards the horizontal part of the kilter line, from $(x(i,j), c(i,j) + \pi_i - \pi_j)$ to $(x(i,j), c(i,j) + \pi_i - \pi_j - \delta)$; similarly, in case (b), the kilter position will be raised by the amount δ.

There will be some arcs which link the labelled and unlabelled nodes which do not fall into the groups defined by the conditions (a) and (b). These will correspond to forward arcs whose kilter position lies to the right of the kilter line and the shaded area of Fig. 4.12, and to reverse arcs which lie to the left of the kilter line and the shaded area of Fig. 4.13. These arcs do not play a part in determining the value of δ.

Once the node multipliers have been changed it will either be possible to label at least one more node, so extending the flow-augmenting chain, or the arc between s and t may have been brought into kilter. If the former, then the process for finding a chain to the node t continues as before. However, the potential capacity of the chain may be affected by the vertical movement of the kilter positions of arcs, and it is often better to start the labelling process afresh from node s. If the latter, a new out-of-kilter arc should be sought.

Eventually, either there will be no out-of-kilter arcs, in which case the optimal flow pattern will have been found, or there will be no arcs between labelled and unlabelled nodes which fall into the groups defined by cases (a) and (b) above. Should this happen, there can be no feasible flow in the network, due to an imbalance of flow constraints.

It is now possible to make a formal statement of the algorithm:

THE OUT-OF-KILTER ALGORITHM

step 0   Assign an arbitrary multiplier π_i to each node i. (These may all be zero.) Assign arbitrary flows to all the arcs, which should satisfy the flow conservation equations at each node, but need not satisfy the bounds on the flows in the arc. Again, zero is a suitable initial value for this.

step 1   Find an arc which is out-of-kilter, linking the nodes s and t. If this arc is out-of-kilter as a result of its flow being to the left of the kilter line, then it will be thought of as being the arc (t,s); if it is to the right of the kilter line, then it will be thought of as being (s,t). (In either case, the arc in question will be brought closer to its kilter line by increasing the flow in the rest of the network from s to t, and ensuring that the flow is conserved at s and t. If there is no such arc, then stop, since all arcs are in kilter, and the optimal flow distribution will have been found.

step 2   Find a flow-augmenting chain from s to t. The chain will contain forward arcs which are out-of-kilter and whose flow is to the left of the kilter line, and arcs which are in kilter, and are on the horizontal portion of the kilter line, to the left of its intersection with the line $x(i,j) = u(i,j)$; the chain will also contain reverse arcs which are out-of-kilter and whose flow is to the right of the kilter line, and arcs which are in kilter and are on the horizontal portion of the kilter line, to the right of its intersection with the line $x(i,j) = l(i,j)$.

The flow-augmenting path will be created by a labelling algorithm, which assigns a two-part label (a_i,b_i) to a node i, wherein a_i is the node preceding i in the chain from s, and b_i is the largest amount of flow which may be sent along the chain without bringing any arc beyond its kilter line. So, b_i is the smaller of the bound on the node a_i, and the amount which may be sent along the link between a_i and i. This is evaluated as:

for a forward arc (a_i,i);

(a)   if $c(a_i,i) + \pi_{a_i} - \pi_i \leqslant 0$, it is $u(a_i,i) - x(a_i,i)$

(b)   if $c(a_i,i) + \pi_{a_i} - \pi_i > 0$, it is $l(a_i,i) - x(a_i,i)$

for a reverse arc (i,a_i):

(a)   if $c(i,a_i) + \pi_i - \pi_{a_i} \geqslant 0$, it is $x(i,a_i) - l(i,a_i)$

(b)   if $c(i,a_i) + \pi_i - \pi_{a_i} < 0$, it is $x(i,a_i) - u(i,a_i)$

If a flow augmenting chain can be found, then the flow along it is changed by the limit on the chain, the same change made to the out-of-kilter arc, and step 1 is repeated; if there is no flow-augmenting chain, then step 3 is performed.

step 3   Examine the arcs which link the labelled and unlabelled nodes. For arcs which are forward (that is, from a labelled node to an unlabelled) and for which $c(i,j) + \pi_i - \pi_j > 0$ and $l_{ij} \leqslant x_{ij} < u_{ij}$ find δ_1 as the smallest value of $c(i,j) + \pi_i - \pi_j$. For arcs which are reverse, and for which $c(i,j) + \pi_i - \pi_j < 0$ and $l_{ij} < x_{ij} \leqslant u_{ij}$ find δ_2 as the smallest value of $|c(i,j) + \pi_i - \pi_j|$. Set δ equal to the smaller of δ_1 and δ_2, and add this to the π-multipliers of all unlabelled nodes. If δ cannot be established by this rule, the out-of-kilter arc may be used to determine it. For (t,s) out-of-kilter, and $x(t,s) = l(t,s)$ then $\delta = |c(t,s) + \pi_t - \pi_s|$. For (s,t) out-of-kilter and $x(s,t) = u(s,t)$ then $\delta = |c(s,t) + \pi_s - \pi_t|$. Repeat step 1.

4.4.1 Worked Example

Consider the problem of finding the minimal cost feasible flow in the network shown (Fig. 4.17). In such a small problem, of course, it is fairly easy to find the optimal flow by inspection, but it is worthwhile examining the process by which the out-of-kilter method finds the same solution.

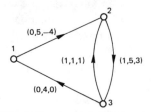

Fig. 4.17 – Arcs labelled (l,u,c).

The algorithm proceeds as follows:

step 0   Set $\pi_1 = \pi_2 = \pi_3 = 0$.
           Set $x(1,2) = x(2,3) = x(3,2) = x(3,1) = 0$.

step 1   Arc $(1,2)$ is out of kilter (Fig. 4.18) so set $t = 1, s = 2$.

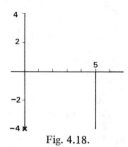

Fig. 4.18.

step 2   Label node $2(1,5)$, node $3(2,\min(5,1-0)) = (2,1)$, node $1(3,\min(4,1)) = (3,1)$.
           Increase the flow in the arcs $(1,2), (2,3), (3,1)$ by 1.
           $x(1,2) = 1, x(2,3) = 1, x(3,1) = 1$.

step 1   Arc (1,2) is out of kilter (Fig. 4.19) so set $t = 1, s = 2$.

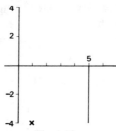

Fig. 4.19.

step 2   Label node 2(1,4), and then no other nodes can be labelled.

step 3   Arc (2,3) is the only forward arc between labelled and unlabelled nodes; it satisfies the other conditions for the algorithm, as $x(2,3) = l(2,3)$ and $c(2,3) + \pi_2 - \pi_3 = 3 > 0$. So $\delta_1 = 3$.

Arc (1,2) is the only reverse arc between the two sets, and it satisfies the other conditions of step 3.

$\delta_2 = 4$; so $\delta = \min(3,4) = 3$

$\pi_1 = 0 + 3 = 3$; $\pi_3 = 0 + 3 = 3$; π_2 is unchanged at 0.

step 1   Arc (1,2) is out of kilter (Fig. 4.20) so set $t = 1, s = 2$.

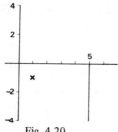

Fig. 4.20.

step 2   Label node 2(1,4), node 3(2,min(4,5−1)) = (2,4),
node 1(3,min(4,4−1)) = (3,3).
Increase the flow in these arcs by 3 units so $x(1,2) = 4$, $x(2,3) = 4$, $x(3,1) = 4$.

step 1   Arc (1,2) is out of kilter (Fig. 4.21) so $t = 1, s = 2$.

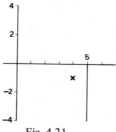

Fig. 4.21.

step 2   Label node $2(1,1)$, node $3(2,\min(1,1))=(2,1)$ node 1 cannot be labelled.
step 3   Arc $(3,1)$ is the only forward arc linking labelled and unlabelled nodes; it does not satisfy the other conditions of step 3 of the algorithm, so $\delta_1 = \infty$; arc $(1,2)$ is the only reverse arc and it does satisfy the other conditions as

$$c(1,2) + \pi_1 - \pi_2 = -1 < 0 \text{ and } l(1,2) < x(1,2) \leqslant u(1,2) \ .$$

So $\delta_2 = 1$, and hence $\delta = \min(\infty,1) = 1$ and $\pi_1 = 3 + 1 = 4$ π_2 and π_3 are unchanged.

step 1   Arc $(1,2)$ is in kilter; arc $(2,3)$ is in kilter; arc $(3,2)$ is out of kilter so $t = 3, s = 2$.
step 2   Label node $2(3,1)$, node $3(2,\min(1,5-4)) = (2,1)$ increase the flow in this chain by 1 unit so $x(3,2) = 1$ and $x(2,3) = 4 + 1 = 5$.
step 1   All arcs are in kilter; stop.

4.4.2 Implementation

There are several problems which are encountered when producing code for the out-of-kilter algorithm. It is evident that an obvious data structure for the network can be defined by a series of inter-related vectors. Each arc will be identified by its two ends, together with the three parameters given to it, its bounds and cost per unit of flow. To the five vectors thus formed, one will be added, corresponding to the flow which has been assigned. Similarly, with each of the nodes, there will be three vectors, two for the labels as flow-augmenting paths are found, and one for the π-multipliers.

   Although it is straightforward to say "this is an out-of-kilter arc", the actual recognition of the state of an arc requires extensive numerical testing, especially in BASIC. The approach which has to be taken in the BASIC program is to identify the six possible ways that an arc may be out of kilter, by finding the position of the arc relative to its kilter line. These six ways are:

(a) $c(i,j) + \pi_i - \pi_j > 0$ $x(i,j) < l(i,j)$

(b) $c(i,j) + \pi_i - \pi_j > 0$ $x(i,j) > l(i,j)$

(c) $c(i,j) + \pi_i - \pi_j = 0$ $x(i,j) < l(i,j)$

(d) $c(i,j) + \pi_i - \pi_j = 0$ $x(i,j) > u(i,j)$

(e) $c(i,j) + \pi_i - \pi_j < 0$ $x(i,j) < u(i,j)$

(f) $c(i,j) + \pi_i - \pi_j < 0$ $x(i,j) > u(i,j)$

In each case, the largest change to the flow which is allowed is calculated. (Again, this change depends on which of these six ways has occurred.)

Once an out-of-kilter arc has been identified, the labelling can be started for the flow-augmenting chain. This follows the same approach as for the maximum-flow method, with arcs being selected as being either forward or reverse, and their capacity for extra flow calculated. This also requires the position of the arc in the kilter diagram to be found, and so there is a much more complex set of statements for finding the new label than in the maximum flow algorithm. The labelling has to be performed in a double loop, since the use of one of the arcs in a chain may mean that one which was earlier found to be unusable can now be used. So, in the labelling sections of the programs, the loops which carry out a scan through the list of arcs are repeated until no more arcs can be used, or until the required chain has been found. In both programs, this repetition is controlled by a variable which is increased when new labels are assigned. Once a chain has been completed, the labels are used to determine the arcs in which flow should be changed, and whether it should be increased or decreased.

The determination of the change to the multipliers, in step 3 of the algorithm, is another situation where logical tests are needed for selecting some arcs which can be used and those which cannot.

In PASCAL, it is slightly easier to perform the tests, and a boolean function has been used to determine whether a given arc is in kilter or not.

```
program prog4p3(input,output);
   const maxnodes = 100;
          inf = 9999;
          maxarcs = 100;
   var    s,t,arc,arcs,node,nodes,plus,minus,pm,
          change,del,arc1,newlabels : integer;
   i,j,x,l,u,c : array[1..maxarcs] of integer;
   a,b,pi : array[1..maxnodes] of integer;
          infeasible,accept,ok : boolean;

   function inkilter(flow,low,high,pc:integer;
##     var changeup,changedown : integer):
          boolean;
   begin
      inkilter := false;
      changeup := 0;
   changedown := 0;
      if (pc>0) then
      begin
 if (flow<low) then changeup := low-flow;
          if (flow=low) then inkilter := true;
 if (flow>low) then changedown := flow-low
      end;
          if (pc=0) then
      begin
 if (flow<low) then changeup := high-flow;
          if ((flow>=low)and(flow<=high)) then
          begin
             inkilter := true;
             changeup := high-flow;
             changedown := flow-low
          end;
```

```
if (flow>high) then changedown := flow-low
  end;
  if (pc<0) then
  begin
if (flow<high) then changeup := high-flow;
    if (flow=high) then inkilter := true;
if (flow>high) then changedown := flow-high
  end
 end; {inkilter}

 function min(a,b : integer) : integer;
 begin
   if (a<b) then min := a else min := b
 end; {min}


 begin
 {Out-of-kilter algorithm
 variables being used (all numerical
##    variables are integers)
 nodes : number of nodes in network
 arcs  : number of arcs
 i,j   : vectors to hold the start node and
##    finish node of each arc
 x     : vector of flows in the arcs
 c     : vector of costs of unit flows in
##    the arcs
 l,u   : vectors of the bounds on the flow
##    in each arc
 a,b   : vector of labels on the nodes
 pi    : vector of pi-multipliers on eah node
 s,t   : ends of the out of kilter arc
 arc   : loop counter and index of the
##    current out of kilter arc
 plus  : permitted increase in flow for the
##    current out of kilter arc
 minus : as for plus,except that it is the
##    decrease
 infeasible : boolean variable indicating
##    whether a value of del has
                 been assigned or not
 del   : change in the pi-multipliers for
##    the unlabelled nodes
 pm    : indicator for whether an arc in the
##    flow augmenting chain is forward
               or reverse
 newlabels : counter of the number of labels
##    asigned in a pass through the arcs
 accept : boolean used to identify arcs
##    which can be used to calculate del
 }
 writeln('Out-of-kilter algorithm');
   repeat
     write('Input number of nodes  and
##    number of arcs  ');
     readln(nodes,arcs);
     ok := ((nodes in [2..maxnodes])and(arcs
##    in [2..maxarcs])and(nodes<arcs));
     if (not ok) then writeln('Error in
##    input;  try again')
   until ok;
   writeln('Input details of the arcs in the
##    following order:');
   writeln('Start node Finish node Lower
##    bound Upper bound Cost');
   writeln('                  S  F  L  U  C');
```

```
(                       ******
                        STEP  0
                        ******
}
   for arc := 1 to arcs do
     repeat
       x[arc] := 0;
       write('Arc number',arc:3,':');
readln(i[arc],j[arc],l[arc],u[arc],c[arc]);
       ok := ((i[arc] in [1..nodes])and(j[arc]
##     in [1..nodes])and(i[arc]<>j[arc])
and(l[arc]>=0)and(u[arc]>=l[arc]));
       if (not ok) then writeln('Error in
##     input:  try again')
     until ok;

   for node := 1 to nodes do  pi[node] := 0;
   infeasible := false;

   for arc := 1 to arcs do
     begin
       while (not inkilter(x[arc],l[arc],
##     u[arc],c[arc]+pi[i[arc]]-pi[j[arc]],
plus,minus)and(not infeasible)) do
       begin
(                       ******
                        STEP  1
                        ******
}

         if (plus>0) then
         begin
           s := j[arc];
           t := i[arc];
           a[s] := arc;
           b[s] := plus
         end
         else
         begin
           t := j[arc];
           s := i[arc];
           a[s] := -arc;
           b[s] := minus
         end;
         repeat
         for node := 1 to nodes do
(                       ******
                        STEP  2
                        ******
}

         if (node <> s) then a[node] := 0;
         repeat
           newlabels := 0;
           for arc1 := 1 to arcs do
if (((a[i[arc1]]=0)and(a[j[arc1]]<>0))or
((a[i[arc1]]<>0)and(a[j[arc1]]=0))) then
           begin
             ok := inkilter(x[arc1],l[arc1],
##     u[arc1],c[arc1]+pi[i[arc1]]-
                    pi[j[arc1]],plus,minus);
if ((a[i[arc1]]<>0)and(plus>0)) then
             begin
               newlabels := newlabels+1;
               a[j[arc1]] := arc1;
b[j[arc1]] := min(b[i[arc1]],plus);
             end
```

```
          else {needed when a[j] has just been set}
          if ((a[j[arc1]]<>0)and(minus>0)) then
                     begin
                        newlabels := newlabels+1 ;
                        a[i[arc1]] := -arc1;
b[i[arc1]] := min(b[j[arc1]],minus);
                     end
                  end
             until ((newlabels=0)or(a[t]<>0));
             if (a[t]<>0) then
             begin
     {                Increase flow in the chain }
                node := t;
                change := b[t];
if (a[s]>0) then x[arc] := x[arc]+change
else x[arc] := x[arc]-change ;
                repeat
                   arc1 := abs(a[node]);
                   if (a[node]>0) then
                   begin
                      node := i[arc1];
                      pm := 1
                   end
                   else
                   begin
                      node := j[arc1];
                      pm := -1
                   end;
                   x[arc1] := x[arc1] +pm*change;
                until (node=s)
             end
             else
             begin
     {               ******
                     STEP 3
                     ******
     }
                del := inf;
                infeasible := true;
                for arc1 := 1 to arcs do

if (((a[i[arc1]]=0)and(a[j[arc1]]<>0))or
((a[i[arc1]]<>0)and(a[j[arc1]]=0))) then
             begin
                accept :=
##    ((x[arc1]<u[arc1])and(x[arc1]>l[arc1]));
      if ((c[arc1]+pi[i[arc1]]-pi[j[arc1]]>0)and
((accept)or(x[arc1]=l[arc1]))) then
                   begin
                      del :=
min(del,c[arc1]+pi[i[arc1]]-pi[j[arc1]]);
                      infeasible := false
                   end;
      if ((c[arc1]+pi[i[arc1]]-pi[j[arc1]]<0)and
((accept)or(x[arc1]=u[arc1]))) then
                   begin
                      del :=
min(del,-(c[arc1]+pi[i[arc1]]-pi[j[arc1]]));
                      infeasible := false
                   end
                end;
             if (del=inf) then
             begin
del := abs(c[arc]+pi[i[arc]]-pi[i[arc]]);
if ((x[arc]>=l[arc])and(x[arc]<=u[arc]))
                   then infeasible:=false
```

```
              end;
                if (infeasible) then
 ##      writeln('There is no feasible flow')
                else
                for node := 1 to nodes do
 if (a[node]=0) then pi[node] := pi[node] +del
            end
           until (infeasible or inkilter(x[arc],
 ##      l[arc],u[arc],c[arc]+pi[i[arc]]-
                  pi[j[arc]],plus,minus))
        end
     end;

     if (not infeasible) then
     begin

        writeln('      Arc     Start    Finish
 ##      Lower     Upper     Cost   Optimal');
        writeln('   Number          Node      Node
 ##      Bound     Bound              Flow');
        for arc := 1 to arcs do
           writeln(arc:8,i[arc]:8,j[arc]:8,
 ##      l[arc]:8,u[arc]:8,c[arc]:8,x[arc]:8);
        writeln;
        writeln('      Node    Pi(n)');
        writeln('   Number');
        for node := 1 to nodes do
           writeln(node:8,pi[node]:8)
     end
 end.
```

```
90 REM PROG4P4
100 DIM X(100),C(100),L(100),U(100),
    P(100),A(100),B(100)
110 DIM I(100),J(100)
120 REM
    *****************************
130 REM              * OUT-OF-KILTER
    ALGORITHM     *
140 REM
    *****************************
150 REM
160 REM              **********
170 REM              * STEP 0 *
180 REM              **********
190 PRINT "OUT-OF-KILTER ALGORITHM"
200 PRINT
210 PRINT "ENTER NUMBER OF NODES,NUMBER
    OF ARCS :";
220 INPUT N,M
230 IF N>100 THEN 2100
240 IF N<1 THEN 2160
250 IF M>100 THEN 2130
260 IF M<1 THEN 2180
270 IF N>M THEN 2200
280 PRINT
290 PRINT "FOR EACH ARC, ENTER DATA IN
    THE FOLLOWING ORDER:"
300 PRINT "START NODE, FINISH NODE,
    LOWER BOUND, UPPER BOUND, COST"
310 FOR M1=1 TO M
320 X(M1)=0
```

```
330 PRINT "ARC NUMBER ";M1;
340 INPUT I(M1),J(M1),L(M1),U(M1),C(M1)
350 IF I(M1)>N THEN 430
360 IF I(M1)<1 THEN 450
370 IF J(M1)>N THEN 470
380 IF J(M1)<1 THEN 490
390 IF I(M1)=J(M1) THEN 510
400 IF L(M1)>U(M1) THEN 530
410 IF L(M1) <0 THEN 550
420 GOTO 570
430 PRINT "START NODE NUMBER TOO LARGE,
        TRY AGAIN"
440 GOTO 330
450 PRINT "START NODE NUMBER TOO SMALL,
        TRY AGAIN"
460 GOTO 330
470 PRINT "FINISH NODE NUMBER TOO
        LARGE,     TRY AGAIN"
480 GOTO 330
490 PRINT "FINISH NODE NUMBER TOO
        SMALL,     TRY AGAIN"
500 GOTO 330
510 PRINT "START NODE AND FINISH NODE
        ARE IDENTICAL,    TRY AGAIN"
520 GOTO 330
530 PRINT "LOWER BOUND EXCEEDS UPPER
        BOUND,    TRY AGAIN"
540 GOTO 330
550 PRINT "LOWER BOUND ON FLOW CANNOT
        BE NEGATIVE,    TRY AGAIN"
560 GOTO 330
570 NEXT M1
580 FOR N1 = 1 TO N
590  P(N1)=0
600 NEXT N1
610 REM               **********
620 REM               * STEP 1 *
630 REM               **********
640 FOR M1 = 1 TO M
650  I1=I(M1)
660  J1=J(M1)
670  C1=C(M1)+P(I1)-P(J1)
680  IF C1 > 0 THEN 750
690  IF C1 < 0 THEN 800
700  B1 = U(M1)-X(M1)
710  IF X(M1) < L(M1) THEN 880
720  B1 = X(M1)-L(M1)
730  IF X(M1) > U(M1) THEN 930
740 GOTO 1920
750  B1 = L(M1)-X(M1)
760  IF X(M1) < L(M1) THEN 880
770  B1 = -B1
780  IF X(M1) > L(M1) THEN 930
790 GOTO 1920
800  B1 = U(M1)-X(M1)
810  IF X(M1) < U(M1) THEN 880
820  B1 = -B1
830  IF X(M1) > U(M1) THEN 930
840 GOTO 1920
850 REM               **********
860 REM               * STEP 2 *
870 REM               **********
880  S=J1
890  T=I1
900  A(S)=M1
910  B(S)=B1
```

```
 920 GOTO 970
 930  S=I1
 940  T=J1
 950  A(S)=-M1
 960  B(S)=B1
 970 FOR N1 = 1 TO N
 980  IF N1=S THEN 1010
 990  A(N1)=0
1000  B(N1)=0
1010 NEXT N1
1020 M2=1
1030 M3=M2
1040 FOR M4 = 1 TO M
1050  I4=I(M4)
1060  J4=J(M4)
1070  IF A(I4) = 0 THEN 1170
1080  IF A(J4) <> 0 THEN 1270
1090  IF X(M4) >= U(M4) THEN 1270
1100  B1 = U(M4)-X(M4)
1110  IF C(M4)+P(I4)-P(J4) <= 0 THEN 1140
1120 IF X(M4) >= L(M4) THEN 1270
1130  B1 = L(M4)-X(M4)
1140  A(J4) = M4
1150  B(J4) = MIN(B1,B1,B1,B(I4))
1160 GOTO 1250
1170 IF A(J4)= 0 THEN 1270
1180 IF X(M4) <= L(M4) THEN 1270
1190  B1 = X(M4) - L(M4)
1200  IF C(M4)+P(I4)-P(J4) >= 0 THEN 1230
1210  IF X(M4) <= U(M4) THEN 1270
1220  B1 = X(M4) - U(M4)
1230  A(I4) = -M4
1240  B(I4) = MIN(B1,B1,B1,B(J4))
1250  M2 = M2 + 1
1260  IF A(T) <> 0 THEN 1300
1270 NEXT M4
1280 IF M3 <> M2 THEN 1030
1290 GOTO 1430
1300 D=B(T)
1310 M4 = A(T)
1320 K4 = ABS(M4)
1330 I4 = I(K4)
1340 J4 = J(K4)
1350 X(K4) = X(K4) + D*SGN(M4)
1360 M5 = I4
1370 IF M4 > 0 THEN 1390
1380 M5 = J4
1390 IF M5 = T THEN 1420
1400 M4 = A(M5)
1410 GOTO 1320
1420 GOTO 650
1430 REM           *********
1440 REM           * STEP 3 *
1450 REM           *********
1460 D = 9999
1470 FOR M4 = 1 TO M
1480  I4 = I(M4)
1490  J4 = J(M4)
1500  D1 = C(M4)+P(I4)-P(J4)
1510 IF A(I4)=0 THEN 1540
1520 IF A(J4) <> 0 THEN 1640
1530 GOTO 1550
1540 IF A(J4) = 0 THEN 1640
1550 IF D1 = 0 THEN 1640
1560 IF D1 > 0 THEN 1610
1570 IF X(M4) =< L(M4) THEN 1640
```

```
1580 IF X(M4) > U(M4) THEN 1640
1590 D = MIN(D,D,D,-D1)
1600 GOTO 1640
1610 IF X(M4) >= U(M4) THEN 1640
1620 IF X(M4) < L(M4) THEN 1640
1630 D = MIN(D,D,D,D1)
1640 NEXT M4
1650 M5=0
1660 IF D<9999 THEN 1720
1670 M5=1
1680 D=C(M1)+P(I1)-P(J1)
1690 D=ABS(D)
1700 IF X(M1)<L(M1) THEN 2060
1710 IF X(M1)>U(M1) THEN 2060
1720 FOR N1 = 1 TO N
1730   IF A(N1) <> 0 THEN 1750
1740   P(N1) = P(N1) + D
1750 NEXT N1
1760 IF M5>0 THEN 650
1770 C1 = C(M1)+P(I1)-P(J1)
1780 IF A(S)>0 THEN 1830
1790 B1=X(M1)-L(M1)
1800 IF C1>=0 THEN 1860
1810 B1 = X(M1)-U(M1)
1820 GOTO 1860
1830 B1=U(M1)-X(M1)
1840 IF C1 <= 0 THEN 1860
1850 B1 = L(M1)-X(M1)
1860 IF B1=B(S) THEN 1030
1870 FOR N1 = 1 TO N
1880    IF A(N1) = 0 THEN 1900
1890    B(N1)=MIN(B1,B1,B1,B(N1))
1900 NEXT N1
1910 GOTO 1030
1920 NEXT M1
1930 PRINT
1940 PRINT "OPTIMUM SOLUTION FOUND"
1950 PRINT
1960 PRINT "LOWER BOUND      UPPER BOUND
       FLOW"
1970 FOR M1 = 1 TO M
1980   PRINT L(M1),U(M1),X(M1)
1990 NEXT M1
2000 PRINT
2010 PRINT "NODE NUMBER      PI()      "
2020 FOR N1 = 1 TO N
2030   PRINT N1,P(N1)
2040 NEXT N1
2050 STOP
2060 PRINT "THERE IS NO FEASIBLE SOLUTION"
2070 PRINT "VALUES FOUND SO FAR"
2080 GOTO 1960
2090 STOP
2100 PRINT "NUMBER OF NODES CANNOT
       EXCEED 100."
2110 PRINT "PLEASE EITHER RE-ENTER, OR
       REDIMENSION THE ARRAYS"
2120 GOTO 210
2130 PRINT "NUMBER OF ARCS CANNOT
       EXCEED 100."
2140 PRINT "PLEASE EITHER RE-ENTER, OR
       REDIMENSION THE ARRAYS"
2150 GOTO 210
2160 PRINT "NUMBER OF NODES MUST BE
       POSITIVE.  TRY AGAIN"
2170 GOTO 210
```

```
2180 PRINT "NUMBER OF ARCS MUST BE
     POSITIVE.  TRY AGAIN"
2190 GOTO 210
2200 PRINT "NUMBER OF ARCS MUST EXCEED
     NUMBER OF NODES."
2210 PRINT "TRY AGAIN"
2220 GOTO 210
```

4.4.3 Applications

The out-of-kilter algorithm may be used for solving several of the problems which have already been encountered in this book, with some slight modifications in some cases to the network. These modifications require extra parameters to be assigned to arcs, and also the extension of networks, by adding one or more new arcs, again with appropriate parameters for each one.

The shortest path problem

Viewed as a network flow problem, the problem of finding the shortest (cheapest) path from node s to node t is simply that of finding the cheapest path for one unit of flow from node s to node t. The arcs have no effective limits on the amount of flow, so the upper bound may be set to ∞, the lower bound set to zero. The cost per unit of flow is set equal to the length of the arc. In order that this one unit can be sent, a new arc (a 'return' arc) is added to the network, with zero cost and bounds which force one unit of flow to pass from t to s. This means that the upper bound and lower bound are set to 1. The out-of-kilter algorithm finds a feasible flow (which means one unit flow in the arc (t,s), and a balancing flow from s to t) and a minimal cost flow (which means that the total cost of the flow from s to t is minimised). The shortest path is identified by the set of arcs with a flow in the solution.

The out-of-kilter algorithm is suitable even for networks in which there are loops with a net negative cost, which was a problem which could not be solved by either of the path algorithms examined. In order that the shortest path from s to t in such a network can be found, the upper bound on all the arcs in the network must be set to 1.

The algorithm can also be used for other shortest path problems, such as the shortest path from node s to all other nodes, and the problem of the second shortest path. In the former case, return arcs must be added to the network from each destination to the node s. These have a unit of flow forced to pass along each, and this is balanced by flow from source to the destinations. The out-of-kilter algorithm cannot be easily used to find all shortest paths in a network, other than by repeated use of it with different sources and sinks. The second shortest path problem can be solved in the same way as in Chapter 3, except that the out-of-kilter algorithm is used in place of Dijkstra's method.

By changing the lower bound on a given arc from zero to one, and applying the out-of-kilter algorithm to the resulting network, it may be possible to find the shortest path from s to t which uses this given arc. However, this may not be guaranteed, as the optimal flow pattern found may be the shortest path from s to t, together with a circulation of flow using the amended arc. However, by exploratory modification of the network, should this latter situation occur, the solution can usually be found to the constrained shortest path problem.

Maximum flow problems
The problem of finding the maximum flow from source s to sink t may also be modified to be solvable by the out-of-kilter algorithm by the addition of a return arc to the original network. However, the objective of this new arc will be to send as large a flow as possible from t to s, in order that as large a flow as possible may be sent from s to t to balance it. So the return arc will be given an infinite upper bound and zero lower bound; its unit cost will be negative, say -1, so that as the flow increases, the total cost decreases. All the other arcs will have zero unit cost, zero lower bound and upper bound set to the capacity of the arc. The out-of-kilter algorithm will find a feasible flow (satisfying all the capacity constraints) at minimal cost (and the only cost contribution is from the flow in the return arc).

Once again, this formulation has numerous alternative features. Arcs can be given lower bounds as well as upper bounds; several sources can be introduced, using a super-source; capacity limits can be given to the nodes.

Transportation, assignment and trans-shipment problems
The classical transportation problem is that represented by Fig. 4.22. A number of factories are to be used to supply shops with some item. Factory i can supply a_i units, and shop j requires b_j units. The cost of transport of one unit from factory i to shop j is c_{ij} units. How should each shop be supplied? This problem is normally solved using an iterative method using a tableau such as that shown in Fig. 4.23. It may also be solved by using the out-of-kilter algorithm. Each possible link from a factory to a shop is represented by an arc (i,j), with infinite upper bound and zero lower bound, with a unit cost set equal to c_{ij}. To this network a new node z is added which is linked to each factory i by an arc (z,i) for which $c(z,i) = 0$, $u(z,i) = l(z,i) = a_i$. Each shop j is linked to z by an arc (j,z) for which $c(j,z) = 0$, $u(j,z) = l(j,z) = b_j$. Then the out-of-kilter algorithm is used to find the minimal-cost feasible flow; the only cost comes from the transport costs, and the feasibility is assured when the amounts flowing from z to each factory are equal to the amounts available there, and when the amounts flowing from each shop to z are equal to the amounts required. Such a formulation of the problem allows bounds to be included on flows between particular factories and shops, a feature which normally adds a complication to the classical transportation algorithm.

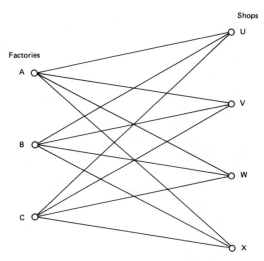

Fig. 4.22.

Shops

	U	V	W	X	Row Total
A	x_{AU} C_{AU}	x_{AV} C_{AV}	x_{AW} C_{AW}	x_{AX} C_{AX}	q_A
B	x_{BU} C_{BU}	x_{BV} C_{BV}	x_{BW} C_{BW}	x_{BX} C_{BX}	q_B
C	x_{CU} CU	x_{CV} C_{CV}	x_{CW} C_{CW}	x_{CX} C_{CX}	q_C
column total	q_U	q_V	q_W	q_X	

Factories (row label)

Fig. 4.23

Several other problems are similar to the transportation problem. The assignment problem and the trans-shipment problem are the best-known of these, and each may be easily formulated as a network problem and solved using the out-of-kilter algorithm.

In the assignment problem, the objective is to form pairs of items in an optimal way. The pairs may be people being given tasks to perform, jobs being assigned to machines, or mixed couples being selected from a list of men and women. Associated with each pair there is a benefit, expressed as an integer, and measured in such a way that low values of benefit are more desirable than high ones. (Such a means of measuring benefit, although not an obvious one, is nonetheless in widespread use. For example, in the classification of degrees, 1 is better than 2, which is better than 3; in the grading of some commodities, the best is recorded as being grade 1, with lesser quality items being given higher numbered grades.) The total benefit from matching several pairs is the sum of the benefits of the individual assignments. Once an assignment has been made, neither item can participate in any other ones; thus, in the case of people with tasks to perform, each person can perform one and only one task, and each task can be done by one and only one person.

Suppose there are n people and m tasks to be performed. The final condition about the assignment problem means that a way should be found to make m and n equal. This may mean that fictitious people have to be assumed to exist, to perform the excess tasks, or fictitious tasks created to occupy the excess people. If there is a benefit c_{ij} associated with person i performing task j ($c_{ij} = 0$ for fictitious people/tasks, and $c_{ij} = \infty$ for impossible assignments), then the assignment problem is that of finding n^2 variables x_{ij} so as to minimise

$$\sum_i \sum_j c_{ij} x_{ij}$$

subject to
$$\sum_i x_{ij} = 1 \text{ (all } j)$$

$$\sum_j x_{ij} = 1 \text{ (all } i)$$

$$x_{ij} = 0 \text{ or } 1 \text{ (all } i \text{ and } j)$$

where the constraints arise from the limits on assignments, and from the common-sense requirement that people either have jobs ($x_{ij} = 1$) or not ($x_{ij} = 0$) and there is no intermediate state.

Such a problem is a special case of the transportation problem, as it has all column totals and row totals equal to 1. It may be solved by the out-of-kilter algorithm in the same way as that problem, by creating a network of nodes representing participants, and arcs representing assignments, together with arcs to force unit flows to and from the items being considered. The traditional approach to solving assignment problems is by means of the "Hungarian algorithm" which lacks the versatility of the out-of-kilter approach.

The trans-shipment problem is a generalisation of the transportation problem. The factories and shops of the latter are retained, but have the added characteristic that they may be used as depots for goods being routed through the network. Additional depots may also exist, where goods may be transferred between vehicles, but which neither supply items nor require any for consumption. Thus it becomes possible, not only to transport goods from factory to shop, but also from factory to factory, from shop to shop, from factory to depot and from depot to shop. With each such route for the transport of goods, there will be a cost corresponding to the transport charge per unit of material carried.

This problem may be converted into a transportation problem by an expansion of the tableau so that there is one row and one column for each factory, shop and depot, with column and row totals so large that every item could possibly flow through the corresponding site. This creates a very large, and frequently cumbersome, tableau. If a network is used, it is straightforward to include new arcs linking the factories, shops and depots as required, with corresponding costs and infinite capacity. The network formulation has the advantage that it does not make any distinction between the roles of nodes, whether they are acting as sources of material, demands for material, or simply as locations where loads may be divided or aggregated.

Non-linear costs on flows

The cost per unit of flow in arcs need not be constant. In some cases, as the flow increases, so does the unit cost. For one such arc, the cost might be equal to the square of the flow (measured in suitable units). Non-linear cost functions such as this can be introduced to the out-of-kilter network so long as the total cost as a function of flow (x) is a convex function $f(x)$. If this is so, then the single arc can be replaced by a succession of arcs each with upper bound of 1, and lower bound 0. The cost for the kth arc will be $f(k) - f(k - 1)$. Since the cost function is convex, $f(k) - f(k-1) > f(k-1) - f(k-2)$, and so the cost of the kth arc is greater than that for the $(k - 1)$th arc; the out-of-kilter algorithm will assign flow to the cheapest arc first, then to the next cheapest, and so on. The flow will cost $f(1) - f(0) = f(1)$ for one unit, $\{f(2) - f(1)\} + \{f(1) - f(0)\} = f(2)$ for two units, and so on, satisfying the original non-linear cost on the arc.

Unfortunately there is no such simple formulation of the problem when the cost function is concave.

Unbalanced flows at nodes

It has been assumed that the flow at every node is conserved. However, the out-of-kilter algorithm never checks that this is so, and thus it is possible to assign initial flows to the network which do not conserve flow at every node. Once initialised in this way, the imbalance will remain throughout all the iterations of the algorithm. Generally this is a nuisance, but it may be used profitably, so long as it is used with caution. In the shortest path problem, one method of solution using the out-of-kilter algorithm would be to initialise the network with a unit flow from s to t along any path, and not use a return arc at all. The algorithm will then redirect this flow as necessary to minimise the total cost.

Sensitivity analysis

The out-of-kilter algorithm lends itself readily to studies of the effect of the addition of arcs, the removal of arcs, and the change of parameters of arcs. Once a solution to a particular network problem has been found, then it may be used as an initial solution for a modified network. If any arcs are out-of-kilter, then the algorithm will correct these and find the optimal solution more rapidly (in most cases) than would happen if the normal initial values were used. One case needs to be treated with caution; if an arc is removed from the network, it must be represented by a dummy arc with zero upper bound in the modified network.

EXERCISES

(4.1)   Use the maximum flow algorithm to find the maximum flow in the network shown (source s, sink t).

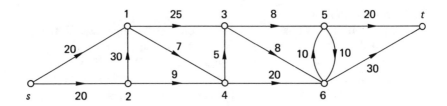

Fig. 4.24.

(4.2)   Solve exercise (4.1) using the out-of-kilter algorithm.

(4.3)   Five jobs have to be assigned to five workers. The efficiency of each possible assignment is measured on a scale from 1 (worst) to 10 (best) as shown in the table below. It is desired to assign the jobs so that the sum of the efficiencies is as large as possible. How can the out-of-kilter algorithm be used to find this assignment?

		Jobs			
	V	W	X	Y	Z
A	8	9	6	4	7
B	6	3	10	6	5
workers C	9	7	1	5	2
D	5	3	7	2	4
E	1	8	4	5	8

(4.4)   In exercise (4.3), suppose the objective were to be changed, so that the assignment is to be made in such a way that all the efficiencies for the assignments are of value N or more, where N is chosen to be as large as possible. How can this be achieved using the out-of-kilter algorithm, if necessary repeatedly?

(4.5) Why, from the point of view of flow in road and rail networks, have bridges been a popular target for bombs in warfare?

(4.6) Using the results of the worked example of the out-of-kilter algorithm, find the optimal flows and π-multipliers when $l(3,2)$ and $u(3,2)$ are increased to 3.

5

Networks for Scheduling Jobs and Activities

This chapter describes a widespread use of networks in industry, for planning and arranging schedules. The networks used in such circumstances indicate the manner in which the constituent activities of some projects are interconnected; analysis of aspects of such networks allows those who are in charge of the project to make efficient use of the resources available for the project.

There are several related techniques which are used for such analysis, and it is only possible here to consider the key features of the main one, known as **Critical Path Analysis.** This was developed in the late 1950s, and gained rapid acceptance in many industries. It has become well established as a valuable tool for enhancing the information available to management and workers involved with a commercial project of some kind. In spite of this wide acceptance, there are many instances where the technique is not used fully, with sad consequences for those concerned.

5.1 CRITICAL PATH ANALYSIS

The range of projects which can be analysed and examined by Critical Path Analysis is enormous, ranging from the planning of a new factory to the preparation of a meal, and from the overhaul of a submarine to the repair of a puncture in a bicycle tyre. In essence, the method identifies those activities from the set which make up the complete project, which are critical for the project, in that they determine how long it will take to complete. In so doing, the inherent flexibility in other activities can be measured, and these can be scheduled to fit in with this flexibility. Additionally, there are further questions about the project which can be answered, such as: 'What activities will be in progress at a given time?' and 'How long can this activity be delayed?'. By simple extensions to the basic method, other, more speculative questions can be answered, such as 'What effect will there be on the project if this activity is completed in half the normal time?'.

The benefits of the Critical Path Analysis method are not confined to numerical information. By identifying a subset of activities whose scheduling is crucial to the completion of the project in the least time, the efforts and skills of managers can be concentrated on such critical parts of the project; hence the name of the technique. Studies of many industrial and commercial enterprises over the last half-century have found that only a small fraction of the activities which go to make up a project require significant intervention by managers. Most activities are completed smoothly, without serious problems. Critical Path Analysis allows a number of crucial activities to be identified in advance, and hence permits the control to be directed to these.

Reference has been made to the set of activities which make up a project, and this indicates one of the key features of an enterprise which enables Critical Path Analysis to be used. In order that the project be completed, each of a set of activities must also be completed. Some of these will depend on earlier activities; others can proceed in parallel with one another.

In order that a project may be represented as a network, it must first be broken down into its constituent activities, and these examined. For each activity, the time which will be taken must be calculated (or estimated as accurately as possible). In addition, the preceding activities for each one must be listed; this will give a set of activities which must have been completed in advance of their successor commencing. The idea of preceding activities may be illustrated by a simple example.

Example of a project split into activities
Suppose that the project 'Write and post a letter' is to be split into its constituent activities. A possible set of these would be:

A	Obtain notepaper
B	Obtain envelope
C	Obtain stamp
D	Write letter
E	Address envelope
F	Affix stamp to envelope
G	Place letter in envelope
H	Seal envelope
I	Post the letter

Activity *D* cannot start until activity *A* has been completed (otherwise what would the letter be written on?). In the same way, activity *I* cannot start until activities *E*, *F* and *H* have been completed. By common-sense arguments of this kind, it is possible to draw up the preceding activities for each activity, thus:

Activity	Predecessor(s)
A	–
B	–
C	–
D	A
E	B
F	B,C
G	B,D
H	G
I	E,F,H

For the first three activities, there are no predecessors, and any of these could be started without any of the others having been commenced. For some of the later activities, it will be seen that only the immediate predecessors have been listed: activity I depends on the completion of the eight activities A–H, but only those three which immediately precede it are given, since these depend, in their turn, on others, and by tracing each one back, it will be found that I does depend on the completion of this set (A–H) of other activities of the project.

In the network representation, each activity will be represented as a direct arc, linking two nodes. These nodes are usually called 'events', and correspond to milestones in the course of the completion of a project. Several arcs may terminate at the same event, which will then correspond to the completion of all the activities which these arcs represent. The arcs which start at a particular node correspond to those activities which depend on the completion of all the activities which terminate at that node. In many cases, it is necessary to 'invent' activities, known as 'dummy activities', whose purpose is to signal the completion of one or more activities. Such an activity takes no time, but is essential for the retention of the logical interdependence of activities in the network, or, as will be explained below, to allow for a convenient representation of the network for analysis.

The network will have one start node, and this will act as the initial node for each activity which has no predecesssors, and one finish node, representing the completion of all the activities, and hence of the project. Besides this, there can be no circuits, since these would lead to a nonsense.

Example

Consider the network described above. There will be one start node, and this will be the first node for *A,B and C.* So these activities could be represented as in Fig. 5.1

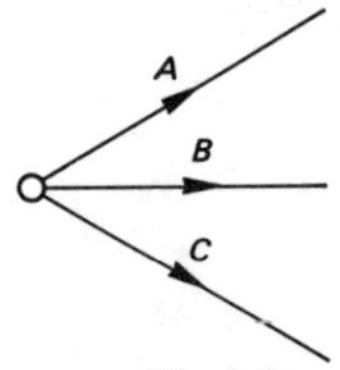

Fig. 5.1.

Activity D depends on the completion of A, and E depends on B. These could be represented thus: (Fig. 5.2)

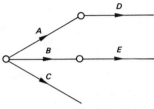

Fig. 5.2.

Activity F depends on the completion of B *and* C. One way of representing this dependence might be to make these two activities terminate in the same event, but this would then imply that E depended on B *and* C, which is not the case. So a dummy activity is needed to show that B is complete and this will terminate at the start node of activity F, thus: (Fig. 5.3)

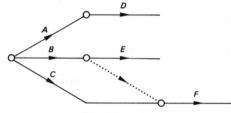

Fig. 5.3.

In the same way, a dummy activity has to be used to signal the finishing of B to allow activity G to commence, thus: (Fig. 5.4)

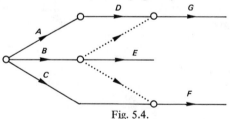

Fig. 5.4.

The network now has 7 of its 9 activities, and it is straightforward to insert the remaining 2, to give a completed network, thus: (Fig. 5.5)

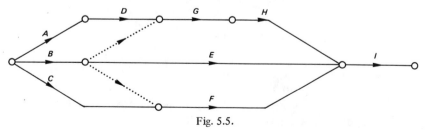

Fig. 5.5.

In general, the process of converting a list of activities into a representation of the network is one which requires several attempts, if an acceptably neat result is desired; the first attempt (and possibly the second and third) will almost invariably be a mess. The process starts, as above, with a start node, and activities are gradually added to it. It is usually easier (as has been the case above) to draw activities without their finish nodes at first, to allow the arcs to be extended or shortened. In most cases, the diagram can be completed in the same way as in the example above. However, where this is not possible, a set of rules for producing a network diagram is as follows:

step 0   Draw the start node, and make this the first node of all the activities which have no predecessors. These may then be drawn in.

step 1   If all the activities have been included, then do step 3.

Otherwise, find an activity which has not been drawn which depends on one (and only one) predecessor, which is already drawn. Draw this in, with its start event the final event of its predecessor. Do step 1 again.

If there are no such activities, then do step 2.

step 2   Find an activity which has not been drawn and which depends on two or more activities which have already been drawn. Draw final events if necessary for each of these activities, and create dummy activities which end on the start node of the activity which has been found, which can now be drawn. Do step 1 again.

step 3   Draw a final event for the project, and make this end node for all the activities which have not already been given end nodes.

When this set of rules is applied to the example of the letter given above, the network which results is Fig. 5.6 (where the dotted arcs represent dummy activities):

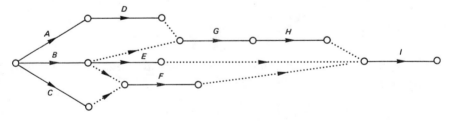

Fig. 5.6.

Apart from (possibly) having the lines which represent arcs crossing over each other, the result of these rules will usually have far too many dummy activities. These can be removed to give a more attractive (and more easily understood) diagram, using two methods.

The first of these is to remove those dummy activities which signal the completion of an activity, which could signal its own completion. Such dummy activities will be characterised by being the only activity to start at a given event node, and can be removed by extending the arc for the activity to the final node of the dummy activity. It may be necessary to retain some of these dummy activities in order that there is only one arc between a given pair of nodes. If there are two or more arcs between some pair, then ambiguity could result when referring to activities, and retention of a dummy activity removes this possibility. However, it is also possible to redefine activities so that one arc represents all the activities which share start and finish events.

The second method eliminates duplicate dummy activities. If some activities have one or more predecessors in common, there will often be a dummy activity for each predecessor and each subsequent activity. This situation may be simplified by introducing an intermediate event which will correspond to the completion of the common subset of predecessor activities.

The application of these rules is illustrated below: ⋏

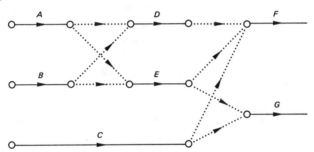

Fig. 5.7 – A portion of a network before removal of unnecessary dummy activities.

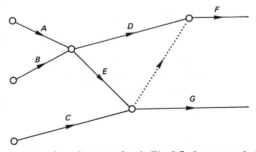

Fig. 5.8 – The same portion of a network as in Fig. 5.7 after removal of unnecessary dummy activities.

Once the superfluous dummy activities have been removed, the network diagram can be tidied up and redrawn. After this has been done, it is convenient to assign numbers to each event, so that the arcs can be readily identified by an ordered parir of nodes, in the same way that arcs in other networks have been identified.

The most widely used method of numbering nodes is to number them consecutively from the start node, number 1, to the project's finish node, using a systematic method. However, in some circumstances, it may be preferable to number events in a systematic way, but with steps of (say) ten, between the numbers. This allows extra activities and events to be added to the network at a later stage without there being the necessity for renumbering the events from scratch. In other circumstances, the numbers on nodes may be assigned to convey some practical information, such as a code to show what reports need to be made when the event is reached. It will be assumed in the discussion which follows that the events have been consecutively numbered, although the adaptation for other conventions is reasonably straightforward. The algorithm to be presented gives numbers to events which obey three rules:

(1) The first event of the project is numbered 1.
(2) If an activity starts at event i and finishes at event j, then i is less than j.
(3) Event i is unique.

In order that these rules can be satisfied, the algorithm numbers the start node, and then repeatedly numbers any unnumbered nodes which can be numbered. A node can be numbered as soon as the start nodes of all the activities for which it is the end node have been numbered. The numbers are assigned consecutively, and where there are several unnumbered events which could be labelled, then an arbitrary choice is made between them.

This method may be presented formally as a simple labelling algorithm, to be used by hand before further analysis of the network is carried out.

5.1.1 The node numbering algorithm

step 0   The network is drawn with no numbers to events.
            The start node is numbered 1. Set $k = 1$.
step 1   Set $k = k + 1$. Select an unnumbered node for which all the preceding activities have a numbered start node. If there is a choice, select any of the possible candidates. Label this node k.
step 2   If there are any unnumbered nodes, then repeat step 1. Otherwise stop.

This algorithm will number all the nodes; if it were otherwise, there would be a stage reached when step 1 could find unnumbered nodes which could not be numbered. Then, for every one of these, there would be a preceding activity which had an unnumbered start node. This 'earlier' unnumbered node must have its own 'earlier' unnumbered node, and by repeating this argument, a sequence of such unnumbered nodes would be found, which would eventually lead to the start node which had been numbered in step 0.

5.1.2 Worked Example

Consider the network shown in Fig. 5.9.

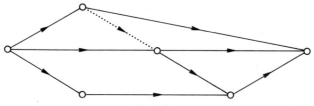

Fig. 5.9.

step 0

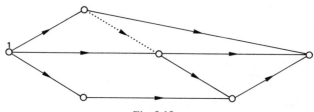

Fig. 5.10.

$k = 1$.

step 1 $k = 2$.

Select arbitrarily *i* or *ii.*

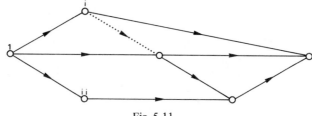

Fig. 5.11.

step 2   Do step 1.

step 1 $k = 3$.

Select arbitrarily *i* or *ii*.

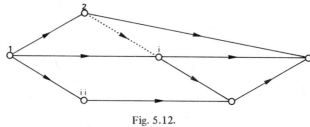

Fig. 5.12.

step 2   Do step 1.

step 1 $k = 4$.

There is no choice as to which node is numbered.

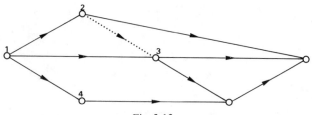

Fig. 5.13.

step 2   Do step 1.

step 1 $k = 5$.

         There is again no choice.

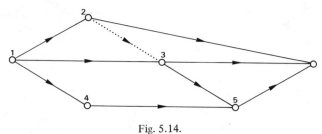

Fig. 5.14.

step 2   Do step 1.

step 1 $k = 6$.

         The final node can be labelled.

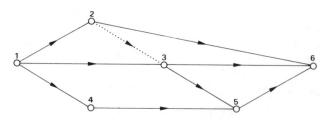

Fig. 5.15.

step 2   Stop.

The labelled network has a use, but only a limited use, for planning a project, since it conveys in a readily understandable form the way that the activities of the project are interrelated. Critical path analysis yields its greatest value when it is used to schedule the use of resources needed and consumed by activities. The commonest resource to be considered is time. Each activity takes a certain amount of time to be completed, and in the light of this requirement, plans must be drawn up in advance. If the project is to be completed as quickly as possible, then the critical activities will be fixed in the times when they can start and finish.

5.1.3 Time analysis of networks

The first step in scheduling activities is to find the duration, $d(i,j)$, for each activity (i,j). These times will determine when each activity can start, as any activity can only start when all its preceding activities have finished, that is, when the event corresponding to the start of the activity has occurred. So, for each event, i, in the network, a label is calculated corresponding to the earliest time when that event can be reached. This label corresponds to the time taken by one series of activities preceding the event, and is the longest such time. It is designated the earliest start time for the event.

The **earliest start time**, ES_i, for event i is the earliest time by which all the activities preceding i can be completed. $ES_1 = 0$ by convention.

The earliest start time for the final event of the project is the shortest possible duration of the project. Earliest start times can be calculated recursively, by a labelling algorithm similar to the one used for numbering the nodes. Starting with the earliest start time of zero on node 1, all nodes which can be labelled are labelled, using the rule:

$ES_j = \max(ES_i + d(i,j))$ where the maximum is taken over all the arcs (i,j) which end at event j.

Since all the nodes have been numbered using the algorithm given earlier, they can be labelled in numerical order. The rule for the labels is derived from the observation that an event can only be reached after all the preceding activities have been completed, and these can only start at the earliest start time of their start event.

This algorithm will eventually label all the nodes; once this has happened, a second label, the latest start time for the event, can be calculated. This may be defined as:

The **latest start time**, LS_i, for event i is the latest time by which all the activities preceding i must have been completed if the project is not to be delayed. The latest start time of the last event of the project is set to be equal to the earliest start time for that event.

The latest start times may also be found recursively, working from the last event of the project backwards to the start. The recursive rule is similar in form to that for the earliest start times, and is:

$LS_i = \min(LS_j - d(i,j))$ where the minimum is taken over all arcs (i,j) which start at event i.

Once again, repeated use of this rule will eventually label all the events; the start of the project will have its early and late start times equal to zero (if the calculation yields any other answer, there has been an error!). Some other events will have both times equal, indicating that these events must not be delayed if the project is to be completed on time.

The times associated with the events can be used to determine when activities should start and finish. In view of the definitions above, the earliest start time of an activity is the earliest start of its start event (since the activities which permit its occurrence cannot be completed any earlier). Similarly, the latest time when an activity may finish without delaying the project is the latest start time of the finish event of that activity. Since the activity takes time to be completed, there are corresponding earliest finish times and latest start times for each. The earliest start time, plus the duration of the activity, yields the earliest finish time; the latest start time is the latest finish time minus the duration.

For many activities in the network, there will be a difference between the earliest and latest start times, and an equal difference between the earliest and latest finish times. This difference is a measure of the flexibility which is possible for scheduling the activity during the project; this flexibility is measured in terms of a 'float time', or, more usually, of 'float'. There are different types of float, corresponding to different limitations which may be imposed on the schedule. Two of these types are in wide use, and two others are encountered in certain scheduling problems. The most commonly used measure is the 'total float' for an activity. This is the difference referred to above, between the earliest start and latest start times for the activity. The total float measures the delay that is possible in an activity without delaying the project, assuming that its start event was reached as early as possible, and that its final event is reached as late as possible. If this amount of float is used, then later activities will be delayed from their earliest starting time, and so will not possess as much flexibility in their schedule. However, the 'free float' is a measure of the time for which an activity may be delayed, without affecting later activities. The free float is the difference between the earliest start of the end event and the earliest finish of the activity. Since the earliest start of the next event is equal to or greater than the earliest finish time, the free float is always non-negative. (In some texts the free float is known as the early free float.)

Two other types of float, which are sometimes met, are derived from the pessimistic assumption that the start event of the activity has been delayed to its latest time. The safety float is the delay which is possible for an activity without delaying the completion of the project, and so is the difference between the latest start time of the final event of the activity and the latest finish time of the activity. The independent float is yet more restrictive; it is the delay that is possible in an activity where the start has been delayed as much as possible, and where later events are to start as early as possible. In some cases it will not be possible to delay the activity at all; then the independent float is zero.

The set of activities which have zero total float (and hence zero values for all the other types of float since the total float is the largest) is the critical path for the project. If any of the activities on the critical path are delayed, then the whole project is delayed.

The different floats may be indicated in a diagram, using horizontal bars to indicate the time taken for the activity, and for the floats which are possible for it. This may be seen in Figs. 5.16 and 5.17, corresponding to the two different possibilities for the independent float.

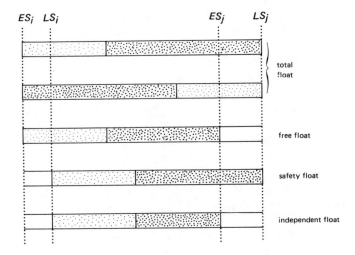

Fig. 5.16.

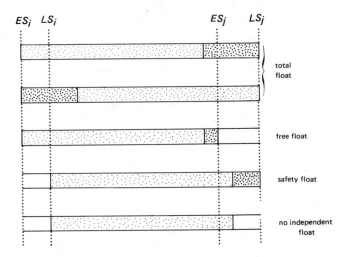

Fig. 5.17.

Calculation of these characteristics of events and activities may be combined into one algorithm, thus:

5.1.4 Algorithm for event and activity characteristics

step 0   For each activity, k, in the network, record $i(k)$ = start event, $j(k)$ = finish event, $d(k)$ = duration.
Sort the activities so that the vector of i's is in ascending order.
For each event, set the earliest start time to zero and the latest start time to infinity.
Set $k = 1$; k records the activity being considered.

step 1   (For the kth activity).
Compare $ES_{j(k)}$ with $ES_{i(k)} + d(k)$, and set $ES_{j(k)}$ to the larger. This gives a new bound for $ES_{j(k)}$. $ES_{i(k)}$ has its true value, since the arcs have been ordered in step 0. Increase k by 1.

step 2   If k is less than or equal to the number of arcs, then do step 1. Otherwise, set k to the number of arcs, and set the latest start time of the final event to be the same as the earliest start time for it. All events now have their earliest start time.

step 3   Compare $LS_{i(k)}$ with $LS_{j(k)} - d(k)$, and set $LS_{i(k)}$ to the smaller. This gives a new bound for $LS_{i(k)}$. Decrease k by 1.

step 4   If k is non-zero, do step 3. Otherwise, do step 5, as all the events have been fully labelled.

step 5   (Calculation of activity characteristics).
For each activity, k, assumed of duration d, linking events i and j:
earliest start (k) = earliest start (i)
latest finish (k) = latest start (j)
earliest finish (k) = earliest start $(k) + d$
latest start (k) = latest finish $(k) - d$
total float (k) = latest finish (k) − earliest finish (k)
free float (k) = earliest start (j) − earliest finish (k)
safety float (k) = latest start (j) − latest start $(i) - d$
independent float (k) = max(0,earliest start (j) − latest start $(i) - d)$

step 6   Stop.

The critical path can be immediately identified by examining the values for the total float, and selecting those activities for which it is zero.

5.1.5 Worked Example

Making a packet jelly
The packet is opened, and the jelly block is cut into pieces. Half a pint of boiling water is measured into a bowl and the pieces are stirred in this until dissolved. The mixture is diluted with cold water, and poured into a mould to set. The empty packet is thrown away.

The activities and durations for this are:

A	Open packet and remove contents	preceded by:	—	Time	0.25 mins
B	Cut jelly block	" "	A	"	1.0 "
C	Boil water and measure it	" "	—	"	2.0 "
D	Dissolve jelly	" "	B,C	"	2.5 "
E	Dilute jelly	" "	D	"	0.5 "
F	Pour into mould and set	" "	E	"	90.0 "
G	Throw packet away	" "	A	"	0.25 "

The network diagram for this project is: ⎿

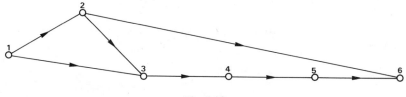

Fig. 5.18.

Following the algorithm, and using the event numbers from the diagram, the activities are:

Activity	i	j	d
A	1	2	0.25
B	2	3	1.0
C	1	3	2.0
D	3	4	2.5
E	4	5	0.5
F	5	6	90.0
G	2	6	0.25

In step 0 of the algorithm, these are ordered, and numbered:

Activity	i	j	d
1	1	2	0.25
2	1	3	2.0
3	2	3	1.0
4	2	6	0.25
5	3	4	2.5
6	4	5	0.5
7	5	6	90.0

Each event is given its two labels, with early start set to 0 and late start to ∞.

step 1 and
step 2

The successive event labels are:

k	event	1	2	3	4	5	6
1		0	0.25	0	0	0	0
2		0	0.25	2.0	0	0	0
3		0	0.25	2.0	0	0	0
4		0	0.25	2.0	0	0	0.5
5		0	0.25	2.0	4.5	0	0.5
6		0	0.25	2.0	4.5	5.0	0.5
7		0	0.25	2.0	4.5	5.0	95.0

The underlined labels are those which may be altered at each stage, as they correspond to the label on $j(k)$. It will be seen that there is no change on any labels when $k = 3$.

step 3 and
step 4

The successive event labels for the latest start are:

k	event	1	2	3	4	5	6
7		∞	∞	∞	∞	5.0	95.0
6		∞	∞	∞	4.5	5.0	95.0
5		∞	∞	2.0	4.5	5.0	95.0
4		∞	94.75	2.0	4.5	5.0	95.0
3		∞	1.0	2.0	4.5	5.0	95.0
2		0	1.0	2.0	4.5	5.0	95.0
1		0	1.0	2.0	4.5	5.0	95.0

step 5   The activity times follow

Activity	early start	early finish	late start	late finish	total float	free float	indept float	safety float
1	0.0	0.25	0.75	1.0	0.75	0.0	0.0	0.75
2	0.0	2.0	0.0	2.0	0.0	0.0	0.0	0.0
3	0.25	1.25	1.0	2.0	0.75	0.75	0.0	0.0
4	0.25	0.5	94.75	95.0	94.25	94.25	93.75	93.75
5	2.0	4.5	2.0	4.5	0.0	0.0	0.0	0.0
6	4.5	5.0	4.5	5.0	0.0	0.0	0.0	0.0
7	5.0	95.0	5.0	95.0	0.0	0.0	0.0	0.0

The critical path is made up of the activities (1.3), (3,4), (4,5), (5,6).

5.1.6 Conventions in drawing networks

There are a number of conventions which are widely, though not universally, observed, when critical path networks are being drawn. The activities are normally drawn with an arc which is made up of a straight line, or a series of straight lines. Often, these are drawn so that at least some of the line is horizontal. And, usually, the direction of the arcs is from left to right; dummy activities are normally shown with a dotted line, and may be vertical. Nodes are normally shown with a small circle, labelled either inside or close by, with the number of the event. The labels for the early and late start times are normally placed in two adjacent squares close to the node. The critical path is sometimes identified, either by using a double line for the activities or by marking each critical activity thus: ———#——— The physical length of an arc has no significance.

These conventions may be ignored where the network is being used to convey information to several groups of people, or individuals. Then different symbols may be used to highlight material which is of especial importance to some of the users of the network — so that one symbol might indicate the responsibility of the purchasing department, another the production department, and so on.

5.1.7 Implementation
The algorithm for finding the attributes of the events and activities of the network is extremely simple to convert into a computer routine. The structure of the network can be conveyed in three vectors, to identify the ends or each activity, and the duration of each one. The event times, and the times and floats for activities can be stored as vectors, calculated exactly as the algorithm specified above has described. Apart from the algorithm, and the input and output sections of the program, the computer program includes two sections which check the data (insofar as this is possible) for inconsistencies. One part ensures that the activities are ordered, and sorts them if necessary to facilitate the labelling later; the other checks that there is only one start node and one finish node for the project, and rejects the network if this condition is not satisfied. The sorting method is the widely used exchange method, which is efficient for small networks; for larger ones, there are alternatives, which are fully discussed in texts on computer searching such as Knuth [8].

```
program prog5p1(input,output);
  const maxnodes = 30;
        maxarcs  = 50;
        inf      = 9999;
  type  actarr   = array[1..maxarcs] of
##     integer;
evearr   = array[1..maxnodes] of integer;
  var   i,j,d,es,ef,ls,lf,ff,tf,ifl,sf :
##     actarr;
        ees,els : evearr;
        events,activits : integer;
        ok : boolean;

  function max(a,b : integer): integer;
  begin
```

```
    if (a<b) then max := b else max := a
end {max};

function min(a,b : integer): integer;
begin
    if (a<b) then min := a else min := b
end {min};

procedure cparead;
var k : integer;
begin
writeln('Critical Path Analysis');
writeln;
repeat
    writeln('How many events, how many
##    activities?');
    readln(events,activits);
    if (events>maxnodes) then writeln('Too
##    many events;  try again');
    if (activits>maxarcs) then writeln('Too
##    many activities;  try again');
    if (activits <1) then writeln('Too few
##    activities;  try again');
    if (events<2) then writeln('Too few
##    events;  try again');
    if (activits<events-1) then writeln('Too
##    many events/too few activities;  try agai
n')
 until((events in [2..maxnodes])and(activits
##    in [1..maxarcs])and(activits>=events-1))
;
 writeln;
 write('Please enter activities with their
##    start event');
 writeln(' finish event and duration');
 for k := 1 to activits do
 repeat
    write('Activity number ',k,' ');
    readln(i[k],j[k],d[k]);
    if (i[k]>=j[k]) then writeln('Start must
##    be less than finish');
    if (d[k]<0) then writeln('Duration must
##    be positive or zero')
 until ((d[k]>=0)and(i[k]<j[k]))
 end { cparead };

procedure cpasort;
var k,l : integer;
    change : boolean;
procedure swap(var a,b : integer);
var c : integer;
begin
    c := a;
    a := b;
    b := c
end { swap } ;

begin
{sorts the set of activits into the order
##    determined by their start events.
    This is needed to facilitate the
##    labelling of events}
change := false;
for k := 1 to activits-1 do
for l := k+1 to activits do
if (i[k]>i[l]) then
```

```
begin
  swap(i[k],i[l]);
  swap(j[k],j[l]);
  swap(d[k],d[l]);
  change := true
end;
if change then writeln('The numbering of
##   the activities has been changed')
end {cpasort};

procedure cpacheck(var ok : boolean);
var check : array[1..maxnodes] of boolean;
    k : integer;
begin
ok := true;
for k := 2 to events do
  check[k] := true;
for k := 1 to activits do
  check[j[k]] := false;
for k := 2 to events do
  if check[k] then
  begin
    writeln('No activities terminate at
##  node ',k);
    ok := false
  end;
for k := 1 to events-1 do
  check[k] := true;
for k := 1 to activits do
  check[i[k]] := false;
for k := 1 to events-1 do
  if check[k] then
  begin
writeln('No activities start at node ',k);
    ok := false
  end;
if not ok then writeln('Network is not
##  complete.  Try again')
end {cpacheck};

procedure getpath;
var k,l : integer;
begin
{               ******
                STEP 0
                ******
}
for l := 1 to events do
begin
  ees[l] := 0;
  els[l] := inf
end;
k := 1;
{               ******
                STEP 1
                ******
}
repeat
ees[j[k]] := max(ees[j[k]],ees[i[k]]+d[k]);
  k := k+1
{               ******
                STEP 2
                ******
}
until (k>activits);
k := activits;
```

```
  els[events] := ees[events];
  {          ******
             STEP 3
             ******
  }
  repeat
els[i[k]] := min(els[i[k]],els[j[k]]-d[k]);
    k := k-1
  {          ******
             STEP 4
             ******
  }
  until (k<1);
  {          ******
             STEP 5
             ******
  }
  for k := 1 to activits do
  begin
    es[k] := ees[i[k]];
    ef[k] := es[k]+d[k];
    lf[k] := els[j[k]];
    ls[k] := lf[k]-d[k];
    tf[k] := lf[k]-es[k]-d[k];
if[k] := max(0,ees[j[k]]-els[i[k]]-d[k]);
    sf[k] := els[j[k]]-els[i[k]]-d[k];
    ff[k] := ees[j[k]]-ees[i[k]]-d[k]
  end
  end {getpath};

  procedure showpath;
  var k : integer;
  begin
writeln('Result of critical path analysis');
  writeln;
  writeln('Activities');
  write('no  start finish duration early
##    early');
  writeln('   late    late total   free indept
##    safety');
  write('     event   event         start
##    finish');
  writeln('  start finish float float  float
##    float');
  for k := 1 to activits do
  begin
    write(k:3,i[k]:6,j[k]:7,d[k]:9,es[k]:6,
##    ef[k]:7);
    write(ls[k]:6,lf[k]:7,tf[k]:6,ff[k]:6,
##    if[k]:7,sf[k]:7);
if (tf[k]=0) then writeln('C') else writeln;
end;
writeln('critical path marked with C');
writeln;
writeln('Times for each event');
writeln;
writeln('no    early  late');
writeln('      start start');
for k :=1 to events do
  writeln(k:5,ees[k]:6,els[k]:6);
writeln
end {showpath };

begin
{ Critical path analysis
```

```
     This program has been written with the main
##      part of the algorithm
in one procedure, getpath, and other
##      procedures for input/output
and verification of the data,     cparead
##      inputs the details of the
network, cpasort arranges the arcs into
##      numerical order, cpacheck
looks for ommissions from the network, and
##      showpath prints out all
the activity and event times.     The modular
##      nature of the program
has not been taken to its fullest extent,
##      since the vectors which
describe the network are global variables,

   Variables used
   (all are integer)
i           : vector of start nodes for
##      activities
 j          : vector of finish nodes for
##      activities
d           : vector of durations for activities
 es         : vector of early start times for
##      activities
 ls         : vector of late start times for
##      activities
 ef         : vector of early finish times for
##      activities
 lf         : vector of late finish times for
##      activities
 ff         : vector of free float times for
##      activities
 tf         : vector of total float times for
##      activities
 ifl        : vector of independent float times
##      for activities
 sf         : vector of safety float times for
##      activities
 ees        : vector of early start times for
##      events
 eef        : vector of early finish times for
##      events
 events   : number of events in network
 activits: number of activities in network
 ok       : boolean variable used to find
##      whether the network is complete
 }
 cparead;
 cpasort;
 cpacheck(ok);
 if ok then
 begin
   getpath;
   showpath
 end
 end,
```

```
90 REM PROG5P2
100 REM CRITICAL PATH ANALYSIS
110 REM
120 REM VARIABLES USED
130 REM
140 REM    I:   VECTOR OF START NODES FOR
    ACTIVITIES
150 REM    J:   VECTOR OF FINISH NODES FOR
    ACTIVITIES
160 REM    D:   VECTOR OF DURATIONS FOR
    ACTIVITIES
170 REM    S:   VECTOR OF EARLY START
    TIMES FOR ACTIVITIES
180 REM    T:   VECTOR OF EARLY FINISH
    TIMES FOR ACTIVITIES
190 REM    U:   VECTOR OF LATE START TIMES
    FOR ACTIVITIES
200 REM    V:   VECTOR OF LATE FINISH
    TIMES FOR ACTIVITES
210 REM    X:   VECTOR OF TOTAL FLOAT
    TIMES FOR ACTIVITIES
220 REM    W:   VECTOR OF FREE FLOAT TIMES
    FOR ACTIVITIES
230 REM    Y:   VECTOR OF INDEPENDENT
    FLOAT TIMES FOR ACTIVITIES
240 REM    Z:   VECTOR OF SAFETY FLOAT
    TIMES FOR ACTIVITIES
250 REM    A:   VECTOR OF EARLY START
    TIMES FOR EVENTS
260 REM    B:   VECTOR OF EARLY FINISH
    TIMES FOR EVENTS
270 REM    N:   NUMBER OF EVENTS IN NETWORK
280 REM    M:   NUMBER OF ACTIVITIES IN NETWORK
290 DIM I(50),J(50),D(50),S(50),T(50),
    U(50),V(50),W(50),X(50),Y(50),Z(50)
300 DIM A(30),B(30)
310 DIM Q(30)
320 PRINT "CRITICAL PATH ANALYSIS"
330 PRINT
340 PRINT "HOW MANY EVENTS, HOW MANY
    ACTIVITIES?";
350 INPUT N,M
360 IF   N>30   THEN 420
370 IF   M>50   THEN 440
380 IF   N<2    THEN 460
390 IF   M<1    THEN 480
400 IF   M<N-1  THEN 500
410 GOTO 520
420 PRINT "TOO MANY EVENTS:   TRY AGAIN"
430 GOTO 340
440 PRINT "TOO MANY ACTIVITIES:   TRY AGAIN"
450 GOTO 340
460 PRINT "TOO FEW EVENTS:   TRY AGAIN"
470 GOTO 340
480 PRINT "TOO FEW ACYIVITIES:   TRY AGAIN"
490 GOTO 340
500 PRINT "TOO MANY EVENTS/TOO FEW
    ACTIVITIES:   TRY AGAIN"
510 GOTO 340
520 PRINT
530 PRINT "PLEASE ENTER ACTIVITIES WITH
    THEIR START EVENT";
540 PRINT " FINISH EVENT AND DURATION"
550 FOR K = 1 TO M
560 PRINT "ACTIVITY NUMBER";K;" :";
570 INPUT I(K),J(K),D(K)
```

```
580 IF I(K)>=J(K) THEN 610
590 IF D(K)<0 THEN 630
600 GOTO 650
610 PRINT "START MUST BE LESS THEN FINISH"
620 GOTO 560
630 PRINT "DURATION MUST BE POSITIVE OR
    ZERO"
640 GOTO 560
650 NEXT K
660 C=0
670 M1=M-1
680 FOR K=1 TO M1
690 K1=K+1
700 FOR L=K1 TO M
710 IF I(K)<=I(L) THEN 820
720 C1=I(K)
730 I(K)=I(L)
740 I(L)=C1
750 C1=J(K)
760 J(K)=J(L)
770 J(L)=C1
780 C1=D(K)
790 D(K)=D(L)
800 D(L)=C1
810 C=1
820 NEXT L
830 NEXT K
840 IF C<1 THEN 860
850 PRINT "THE NUMBERING OF THE
    ACTIVITIES HAS BEEN CHANGED"
860 C=1
870 FOR K= 2 TO N
880 Q(K)=1
890 NEXT K
900 FOR K = 1 TO M
910 C1=J(K)
920 Q(C1)=0
930 NEXT K
940 FOR K=2TO N
950 IF Q(K)<1 THEN 980
960 PRINT "NO ACTIVITIES TERMINATE AT
    NODE ";K
970 C=0
980 NEXT K
990 N1=N-1
1000 FOR K=1 TO N1
1010 Q(K)=1
1020 NEXT K
1030 FOR K=1 TO M
1040 C1 = I(K)
1050 Q(C1)=0
1060 NEXT K
1070 FOR K= 1 TO N1
1080 IF Q(K)<1 THEN 1110
1090 PRINT "NO ACTIVITIES START AT NODE";K
1100 C=0
1110 NEXT K
1120 IF C>0 THEN 1180
1130 PRINT "NETWPRK IS NOT COMPLETE:
     TRY AGAIN"
1140 GOTO 320
1150 REM        ******
1160 REM        STEP 0
1170 REM        ******
1180 FOR L=1 TO N
1190 A(L)=0
```

```
1200 B(L)=9999
1210 NEXT L
1220 K=1
1230 REM            ******
1240 REM            STEP 1
1250 REM            ******
1260 C1=I(K)
1270 C2=J(K)
1280 A(C2)=MAX(A(C2),A(C2),A(C2),A(C1)+D(K))
1290 K=K+1
1300 REM            ******
1310 REM            STEP 2
1320 REM            ******
1330 IF K<=M THEN 1260
1340 K=M
1350 B(N)=A(N)
1360 REM            ******
1370 REM            STEP 3
1380 REM            ******
1390 C1=I(K)
1400 C2=J(K)
1410 B(C1)=MIN(B(C1),B(C1),B(C1),B(C2)-D(K))
1420 K = K-1
1430 REM            ******
1440 REM            STEP 4
1450 REM            ******
1460 IF K>=1 THEN 1390
1470 REM            ******
1480 REM            STEP 5
1490 REM            ******
1500 FOR K=1 TO M
1510 C1=I(K)
1520 C2=J(K)
1530 S(K)=A(C1)
1540 T(K)=S(K)+D(K)
1550 V(K)=B(C2)
1560 U(K)=V(K)-D(K)
1570 X(K)=V(K)-S(K)-D(K)
1580 Y(K)=MAX(0,0,0,A(C2)-B(C1)-D(K))
1590 Z(K)=B(C2)-A(C1)-D(K)
1600 W(K)=A(C2)-A(C1)-D(K)
1610 NEXT K
1620 PRINT "RESULT OF CRITICAL PATH
     ANALYSIS"
1630 PRINT
1640 PRINT "ACTIVITIES"
1650 PRINT "NO  START FINISH DURATION
     EARLY EARLY";
1660 PRINT " LATE LATE TOTAL  FREE
     INDEPT SAFETY"
1670 PRINT "     EVENT  EVENT
     STARTFINISH";
1680 PRINT " STARTSTART FLOAT FLOAT
     FLOAT FLOAT"
1690 FOR K=1TO M
1700 PRINT
     K;I(K);J(K);D(K);S(K);T(K);U(K);V(K);X(K);W(K);
1710 PRINT Y(K);Z(K);
1720 IF X(K)<1 THEN 1750
1730 PRINT
1740 GOTO 1760
1750 PRINT "C"
1760 NEXT K
1770 PRINT "CRITICAL PATH MARKED WITH C"
1780 PRINT
1790 PRINT "TIMES FOR EACH EVENT"
```

```
1800 PRINT
1810 PRINT "NO      EARLY   LATE"
1820 PRINT "        START   START"
1830 FOR K=1 TO N
1840 PRINT K;A(K);B(K)
1850 NEXT K
1860 PRINT
1870 STOP
```

5.1.8 Scheduling the project

The construction of the table of start and finish times, and the calculation of the floats for each activity has identified the flexibility which is possible for the starting of each of these. It has, however, concentrated on the time that each activity takes; other resources are important for the management of a successful project. So, a building company, working on a new housing estate, will be concerned about the adequate provision of each of a number of specialist craftsmen — it would be disastrous if all the houses were to need (say) plumbers at the same time if this situation were immediately followed by a period when there was no need for plumbing work. The company may also be concerned about an adequate flow of money from purchasers — to pay for further developments. Each resource, whether of particular tradespeople or of finance, will have an effect on the scheduling of the activities within the timespan available for them.

The concept of resource smoothing has been devised to assist in the allocation of activities within the range of time which is open to each one. Suppose that each activity requires that units of a particular resource be available for the whole of the duration of the activity. So, one activity might require 15 workers, another only 4. The total requirements for this skill may be plotted against time by a bar chart such as that of Fig. 5.19. This shows that for the first three time units, 7 workers were required, then for the next two, 12 workers were needed, and so on. The maximum ever needed was 19, for the time unit between

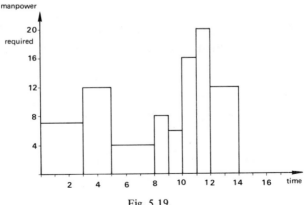

Fig. 5.19.

11 and 12. Such bar charts may be drawn for any critical path network, assuming that a start time has been fixed for each activity, and for all types of resources being considered. By use of resource smoothing, it is frequently possible to rationalise the use of resources over time to smooth the peaks and troughs of demand.

The methods used tend to be heuristic, rather than exact, since for most project networks, there will be too many possible schedules to evaluate each one in a reasonable length of time. Even where there is only one resource to be considered, there are three situations in which resource smoothing may be introduced. The first is for scheduling activities so that the project is completed as quickly as possible, which means that all the activities on the critical path are fixed in time. The second is for scheduling activities so that the maximum use of the resource is minimised (to answer the question 'How should this set of activities be performed, given a constant size workforce of minimal size?'). The third is a combination of the other two, where there is a cost associated with the maximum resource usage and with the duration of the project; so, this last case is important when trying to strike a balance between the minimal duration and minimal workforce.

In each case, the procedure is similar. The activities are scheduled initially to start at their earliest start time. This will yield a bar-chart of resource versus time similar to Fig. 5.19. The different activities are identified on this, and the critical activities are marked. By eye, those activities which lead to peaks of resource demand can be identified. These can be delayed, and the effect on the resource use noted. Hopefully, in most cases, a slight delay in a non-critical activity will decrease the need for the resource. However, there will be a limit, determined by its total float, which will prevent such an activity being delayed by too much. This process, of visually moving the blocks of resource demand around on the diagram will frequently solve the first of the problems above. The second and third can be considered as extensions of the method, but the critical activities can now be moved around as well.

Example

step 1   Consider the very simple network shown below (Fig. 5.20). Both activity (1,2) and activity (1,3) require one unit of a particular resource.

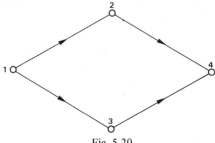

Fig. 5.20.

If these two activities are scheduled to start at their earliest start times, then the maximal usage at any time is two units, as shown in Fig. 5.21.

start	finish	duration	resource need
1	2	2	1
1	3	4	1
2	4	4	0
3	4	6	0

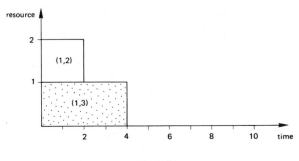

Fig. 5.21.

The non-critical activity (1,2) may be delayed by 4 units, so that it starts when the resource has been released by (1,3), to give a resource usage thus (Fig. 5.22).

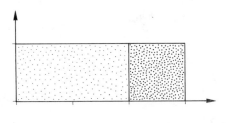

Fig. 5.22.

step 2   However, if the duration of (1,2) were to be 3 time units instead of 2, it would be impossible to delay its start more than 3 time units, so that all schedules would need 2 units of resource for the project to be completed within the minimal time of 10 units. A modest delay of 1 unit, on the other hand, would mean that the project could be completed without ever needing more than one unit of resource.

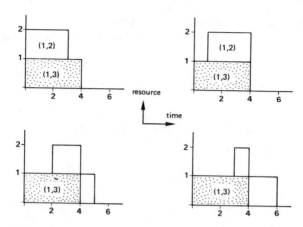

Fig. 5.23 – Possible schedules if (1,2) takes 3 time units.

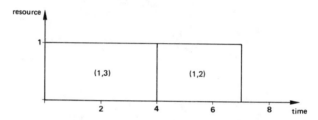

Fig. 5.24 – A schedule which extends the project, but reduces the use of the resource.

step 3   In the network below, the non-critical activity (4,5), which has a high demand for workers, prevents a smooth allocation of resources unless the project is delayed. (Figs. 5.25 – 5.28).

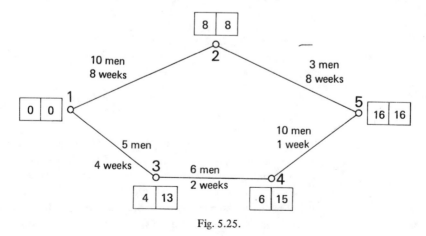

Fig. 5.25.

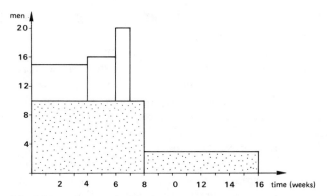

Fig. 5.26 – All activities at earliest start time need 20 men.

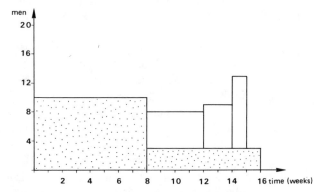

Fig. 5.27 – All non-critical activities smoothed. 13 men needed.

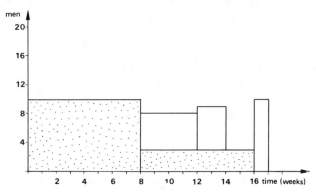

Fig. 5.28 – Activity (4,5) delayed, 10 men needed.

There are several computer programs available commercially for resource smoothing, which provide a large number of options for smoothing more than one resource and including limited delays on activities. These are outside the scope of this book.

5.2 PROJECT EVALUATION AND REVIEW TECHNIQUE (PERT)

The methods described already assume that all the durations of activities are known precisely in advance. Such an assumption is not always justified in practice. In many projects, some or even all of the activities may have a duration which depends on many random events. Then, the duration which is available at the time of drawing up a network diagram can only be an estimate, with some measure of error attached to it. It is easy to see ways in which an activity may be delayed. Bad weather may prevent construction work, or delay transport; a breakdown may affect the completion of a production run; raw materials may be held up. On the other hand, an activity may be completed more quickly than expected. A production process may run without any delays; a new recruit may learn his job more quickly than expected, and so speed up a piece of work; the R & D department may supply answers more quickly than they expected. Uncertainties in the duration of activities arise from these and numerous other circumstances.

In order that such difficulties may be effectively included in the analysis of a project, the technique known as **Project Evaluation and Review Technique** (PERT for short) was developed at about the same time as critical path analysis methods were being studied. The two are very similar in many respects, as PERT also gives rise to network diagrams of the same kind as those given above, and one measure of the duration of the project is provided by the duration of the activities on a critical path. But, instead of using only one figure for the duration of an activity there are three. These correspond to three different ideas about the likely time which an activity may take. There is an **optimistic** time, usually denoted a, which is the shortest time which the activity would ever take, if it were to be repeatedly performed under the same conditions. There is a **pessimistic** time, often denoted b, which is the longest time the project would take, once again given repeated performance under identical conditions. And there is, between the two, a **most likely** time, denoted m, whose title is self-explanatory. Together, these roughly characterise the distribution of a random variable for the duration, t, of the activity; for this, m is the mode (not the mean necessarily), and the range is $a \leqslant t \leqslant b$ (There have been some suggestions that a and b be defined to be the values which correspond to the 5% and the 95% points of the distribution. In other words, on one repetition in twenty, t would be less than a; and on one repetition in twenty, t would be greater than b. This idea has several benefits, but has not been widely accepted, yet.)

For the purposes of analysis, other statistics about the random variable t are needed. These are estimated from a, b and m, whose principal virtue is that they can easily be explained in questions to a manager . . . 'If everything went wrong, how long would this activity take?' . . . The mean and the variance are the two most useful statistics for the analyst. The mean is calculated according to the expression:

mean $= (a + 4m + b)/6$

which corresponds to (among others) t having a beta distribution. The variance is found from the expression:

$$\text{variance} = (b-a)^2/36$$

This corresponds to the fairly widely occurring situation that the standard deviation (the square root of the variance) is close to the range divided by 6, for continuously distributed unimodal random variables. (If the alternative definitions of a and b are used, then these expressions are slightly changed.)

The mean duration of the activity is used in the network to find the time characteristics of the activities and of the project. The critical path is found, and the analysis is concentrated on this. The mean duration of the project will be the sum of the mean durations of the activities on the critical path, assuming, as is reasonable, that these are independent of one another. In addition to calculating event times for all the events of the project, it is usual to calculate the variance of these early start times. The variance is found using the assumption that the durations of activities are independent random variables. The variance of the early start time of an event is the sum of the variances of all the activities which determined that early start time. If two or more sets of activities gave the same early start time, then the variance associated with the event is the largest such sum of variances. So the early start variance for each event can be calculated at the same time as the early start itself is being evaluated, recursively, and eventually the variance of the whole project will be found as the variance associated with the final node. So, the mean duration of the project, and the variance of this, can be found, and used to answer questions about the range of time for the project to take. In most cases, the distribution of the duration of the project may be assumed to be normal, with the mean and variance as above. So, it is possible to answer questions such as: what is the probability that this project will take 10% less time than forecast?

5.2.1 Worked Example

For a particular project, shown in Fig. 5.29, the activities have characteristic times as below: what is the expected duration of the project, and how likely is it that it will last more than three days more than this?

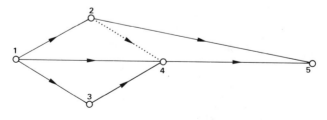

Fig. 5.29.

i	j	a	b	m	mean	variance
1	2	17	25	20	20.33	1.7778
1	3	5	13	11	10.33	1.7778
1	4	10	15	14	13.5	0.69
2	4	0	0	0	0.0	0.0
2	5	6	9	7	7.16	0.25
3	4	8	8	8	8.0	0.0
4	5	14	18	16	16.0	0.4444

The expected duration of the project can be found by the labelling algorithm, which identifies the path (1,2), (2,4), (4,5) as being critical; the expected duration is therefore:

$$20.33 + 0.0 + 16.0 = 36.33 \text{ days.}$$

The variance of the duration of this path is

$$1.7778 + 0.0 + 0.4444 = 2.2222 \text{ (days)}^2$$

and so the standard deviation is $\sqrt{2.2222} = 1.49$ days.

The probability of exceeding the expected duration of the project by 3 days is equal to the probability that a standardised unit normal random variable is more than $3/1.49 = 2.01$, which is 0.0222. So the probability that the project takes more than 39.33 days is 0.0222.

5.2.2 Implementation
No specific algorithm for PERT networks has been given, as the method shares much in common with critical path analysis. The programs below, therefore, have many features which are common to both methods, but differ in regard to the input and output. The programs ask for the three characteristic times for each activity, and then calculate the mean duration according to the expression above. This mean duration is used to label the events, and to find the critical path through the network. The variances of all the activities on this path are found, and from these, the variance of the project duration is calculated.

```
program prog5p3(input,output);
   const maxnodes = 30;
         maxarcs  = 50;
         inf      = 9999;
   type  actarr   = array[1..maxarcs] of
##       integer;
   revearr  = array[1..maxnodes] of real;
   ractarr  = array[1..maxarcs] of real;
   var    i,j : actarr;
          a,b,m,mean,vari,es,ef,ls,lf,ff,tf,ifl,
##     sf : ractarr;
          evar,ees,els : revearr;
          events,activits : integer;
          old : real;
          ok : boolean;
```

```
function rmax(a,b : real): real;
begin
  if (a<b) then rmax := b else rmax := a
end {rmax};

function rmin(a,b : real): real;
begin
  if (a<b) then rmin := a else rmin := b
end {rmin};

procedure pertread;
var K : integer;
begin
writeln('P.E.R.T. (Program Evaluation and
##   Review Technique)');
  writeln;
  repeat
    writeln('How many events, how many
##    activities?');
    readln(events,activits);
    if (events>maxnodes) then
writeln('Too many events;  try again');
    if (activits>maxarcs) then
writeln('Too many activities;  try again');
    if (activits <1) then
writeln('Too few activities;  try again');
    if (events<2) then
writeln('Too few events;  try again');
    if (activits<events-1) then
        writeln('Too many events/too few
##    activities;  try again')
  until((events in [2..maxnodes])
        and(activits in [1..maxarcs])
        and(activits>=events-1));
  writeln;
  write('Please enter activities with their
##   start event');
  writeln(' finish event and times a,b,m');
  for K := 1 to activits do
  repeat
    write('Activity number ',K,'  ');
    readln(i[K],j[K],a[K],b[K],m[K]);
    mean[K] := (a[K]+4*m[K]+b[K])/6.0;
    vari[K]  := (b[K]-a[K])*(b[K]-a[K])/36.0;
    if (i[K]>=j[K]) then
writeln('Start must be less than finish');
    if ((a[K]<0)or(b[K]<0)or(m[K]<0)) then
        writeln('All times must be positive or
##   zero');
    if ((a[K]>m[K])or(m[K]>b[K])) then
writeln('Times must be ordered a<=m<=b')
  until ((a[K]>=0)and(b[K]>=m[K])
        and(m[K]>=a[K])and(i[K]<j[K]))
  end { pertread };

procedure pertsort;
var K,l : integer;
    change : boolean;
procedure swapr(var a,b : real);
var c : real;
begin
  c := a;
  a := b;
  b := c
end { swapr } ;
```

```
procedure swap(var a,b : integer);
var c : integer;
begin
   c := a;
   a := b;
   b := c
end { swapr } ;

begin
{sorts the set of activities into the order
##    determined by their start events.
   This is needed to facilitate the
##    labelling of events}
change := false;
for K := 1 to activits-1 do
for l := k+1 to activits do
if (i[k]>i[l]) then
begin
   swap(i[k],i[l]);
   swap(j[k],j[l]);
   swapr(a[k],a[l]);
   swapr(b[k],b[l]);
   swapr(m[k],m[l]);
   swapr(mean[k],mean[l]);
   swapr(vari[k],vari[l]);
   change := true
end;
if change then
   writeln('The numbering of the activities
##    has been changed')
end {pertsort};

procedure pertcheck(var ok : boolean);
var check : array[1..maxnodes] of boolean;
    k : integer;
begin
ok := true;
for K := 2 to events do
   check[k] := true;
for K := 1 to activits do
   check[j[k]] := false;
for K := 2 to events do
   if check[k] then
   begin
      writeln('No activities terminate at
##    node ',k);
      ok := false
   end;
for K := 1 to events-1 do
   check[k] := true;
for K := 1 to activits do
   check[i[k]] := false;
for K := 1 to events-1 do
   if check[k] then
   begin
writeln('No activities start at node ',k);
      ok := false
   end;
if not ok then
   writeln('Network is not complete.  Try
##    again')
end {pertcheck};

procedure getpert;
var k,l : integer;
begin
```

```
{               ******
                STEP 0
                ******
}
for l := 1 to events do
begin
  ees[l] := 0;
  evar[l] := 0;
  els[l] := inf
end;
K := 1;
{               ******
                STEP 1
                ******
}
repeat
  old := ees[j[k]];
  ees[j[k]] := rmax(old,ees[i[k]]+mean[k]);
  if (ees[j[k]]>old) then evar[j[k]] :=
##    evar[i[k]] + vari[k];
  if (ees[j[k]]=old) then evar[j[k]] :=
##    rmax(evar[j[k]],evar[i[k]]+vari[k]);
  k := K+1
{               ******
                STEP 2
                ******
}
until (k>activits);
K := activits;
els[events] := ees[events];
{               ******
                STEP 3
                ******
}
repeat
  els[i[k]] := rmin(els[i[k]],
##    els[j[k]]-mean[k]);
  k := K-1
{               ******
                STEP 4
                ******
}
until (K<1);
{               ******
                STEP 5
                ******
}
for k := 1 to activits do
begin
  es[k] := ees[i[k]];
  ef[k] := es[k]+mean[k];
  lf[k] := els[j[k]];
  ls[k] := lf[k]-mean[k];
  tf[k] := lf[k]-es[k]-mean[k];
  if[k] := rmax(0,
##    ees[j[k]]-els[i[k]]-mean[k]);
  sf[k] := els[j[k]]-els[i[k]]-mean[k];
  ff[k] := ees[j[k]]-ees[i[k]]-mean[k]
end
end {getpert};

procedure pertpath;
var k : integer;
begin
writeln('Result of PERT analysis');
writeln;
```

```
  writeln('Activities');
  writeln('no   start finish     duration
##    optimistic pessimistic most likely');
  writeln('   event  event      (mean)
##    time      time        time');
  for k := 1 to activits do
    writeln(k:3,i[k]:6,j[k]:7,mean[k]:12:4,
##    a[k]:12:4,b[k]:12:4,m[k]:12:4);
  writeln;
writeln('Activity start and finish times');
  writeln;
  writeln('  no       early       early
##    late    late');
  writeln('             start       finish
##    start      finish');
  for k := 1 to activits do
    writeln(k:3,es[k]:12:4,ef[k]:12:4,
##    ls[k]:12:4,lf[k]:12:4);
  writeln;
  writeln('Floats for the activities');
  writeln;
  writeln('  no        total        free
##    indept      safety');
  writeln('            float        float
##    float      float');
  for k := 1 to activits do
  begin
    write(k:3,tf[k]:12:4,ff[k]:12:4,
##    if[k]:12:4,sf[k]:12:4);
    if (tf[k]<=0.0001) then
      writeln('C') else writeln
  end;
  writeln('critical path marked with C');
  writeln;
  writeln('Times for each event');
  writeln;
  writeln('no        early        late
##    variance');
  writeln('           start        start');
  for k :=1 to events do
    writeln(k:3,ees[k]:12:4,els[k]:12:4,
##    evar[k]:12:4);
  writeln;
  writeln('The expected duration of the
##    project is',ees[events]:12:4);
  writeln('This time has variance ',
##    evar[events]:12:4);
  evar[events] := sqrt(evar[events]);
  writeln('and standard deviation ',
##    evar[events]:12:4)
  end {pertpath };


  begin
  { PERT path analysis

  This program has been written with the main
##    part of the algorithm
  in one procedure, getpert, and other
##    procedures for input/output
  and verification of the data,   pertread
##    inputs the details of the
  network, pertsort arranges the arcs into
##    numerical order, pertcheck
  looks for ommisions from the network, and
##    pertpath prints out all
  the activity and event times,   The modular
```

```
##    nature of the program
 has not been taken to its fullest extent,
##    since the vectors which
 describe the network are global variables,

 Variables used
 ( integer)
 i          : vector of start nodes for
##    activities
 j          : vector of finish nodes for
##    activities
 (real)
 mean       : vector of (assumed mean)
##    durations for activities
 a          : vector of optimistic times for
##    activities
 b          : vector of pessimistic times for
##    activities
 m          : vector of most likely times for
##    activities
 vari       : vector of (assumed) varinces of
##    duration of activities
 es         : vector of early start times for
##    activities
 ls         : vector of late start times for
##    activities
 ef         : vector of early finish times for
##    activities
 lf         : vector of late finish times for
##    activities
 ff         : vector of free float times for
##    activities
 tf         : vector of total float times for
##    activities
 ifl        : vector of independent float times
##    for activities
 sf         : vector of safety float times for
##    activities
 ees        : vector of early start times for
##    events
 eef        : vector of early finish times for
##    events
 evar       : vector of variances of early
##    times for events
 events : number of events in network
 activits: number of activities in network
 ok         : boolean variable used to find
##    whether the network is complete
 }
 pertread;
 pertsort;
 pertcheck(ok);
 if ok then
 begin
   getpert;
   pertpath
 end
 end,
```

```
90 REM PROG5P4
100 REM P.E.R.T.
110 DIM I(30),J(30),A(30),B(30),M(30),
    N(30),V(30),C(30)
120 DIM D(30),E(30),F(30),G(30),H(30),
    O(30),P(30)
130 DIM R(20),S(20),T(20),Q(20)
140 REM PERT. VARIABLES USED
150 REM I : START NODE
160 REM J : FINISH NODE
170 REM A : OPTIMISTIC TIME
180 REM B : PESSIMISTIC TIME
190 REM M : MOST LIKELY TIME
200 REM N : MEAN TIME
210 REM V : VARIANCE OF TIME
220 REM C : EARLY START TIME
230 REM D : EARLY FINISH TIME
240 REM E : LATE START TIME
250 REM F : LATE FINISH TIME
260 REM G : FREE FLOAT
270 REM H : TOTAL FLOAT
280 REM O : INDEPENDENT FLOAT
290 REM P : SAFETY FLOAT
300 REM R : EVENT VARIANCE
310 REM S : EVENT EARLY START
320 REM T : EVENT LATE START
330 REM E9: MAXIMUM NUMBER OF EVENTS
340 REM N9 : MAXIMUM NUMBER OF ACTIVITIES
350 REM E1 : NUMBER OF EVENTS
360 REM N1 : NUMBER OF ACTIVITIES
370 REM T9 : INFINITY
380 REM T5 : ZERO/ONE VARIABLE USED FOR
    TESTS
390 REM K,L: LOOP COUNTERS
400 REM E2,N2,K1,T6 : TEMPORARY VARIABLES
410 REM Q : USED TO CHECK VALID DATA
420 PRINT "P.E.R.T. PROGRAM EVALUATION
    AND REVIEW TECHNIQUE"
430 E9=20
440 N9=30
450 T9=9999
460 PRINT
470 PRINT "HOW MANY EVENTS, HOW MANY
    ACTIVITIES?"
480 INPUT E1,N1
490 T5=0
500 IF E1>E9 GOSUB 2210
510 IF N1>N9 GOSUB 2240
520 IF E1<=0 GOSUB 2270
530 IF N1<=0 GOSUB 2300
540 IF N1<E1-1 GOSUB 2330
550 IF T5>0 THEN 470
560 PRINT
570 PRINT"ENTER ACTIVITIES WITH THEIR
    START EVENT,FINISH EVENT"
580 PRINT"AND THE TIMES A,B,M"
590 FOR K=1 TO N1
600 PRINT "ACTIVITY NUMBER ";K;" ";
610 INPUT I(K),J(K),A(K),B(K),M(K)
620 T5 = 0
630 IF I(K)>=J(K) GOSUB 2360
640 IF A(K)<0 GOSUB 2390
650 IF B(K)<0 GOSUB 2390
660 IF M(K)<0 GOSUB 2390
670 IF A(K)>M(K) GOSUB 2420
680 IF M(K)>B(K) GOSUB 2420
```

```
690  IF T5>0 THEN 600
700  N(K)=(A(K)+B(K)+4*M(K))/6.0
710  V(K)=(B(K)-A(K))*(B(K)-A(K))/36.0
720  NEXT K
730  N2=N1-1
740  T5=0
750  FOR K=1 TO N2
760  K1=K+1
770  FOR L=K1 TO N1
780  IF I(K)<=I(L) THEN 1010
790  T6=I(K)
800  I(K)=I(L)
810  I(L)=T6
820  T6=J(K)
830  J(K)=J(L)
840  J(L)=T6
850  T6=A(K)
860  A(K)=A(L)
870  A(L)=T6
880  T6=B(K)
890  B(K)=B(L)
900  B(L)=T6
910  T6=M(K)
920  M(K)=M(L)
930  M(L)=T6
940  T6=N(K)
950  N(K)=N(L)
960  N(L)=T6
970  T6=V(K)
980  V(K)=V(L)
990  V(L)=T6
1000 T5=1
1010 NEXT L
1020 NEXT K
1030 IF T5=0 THEN 1050
1040 PRINT "THE NUMBERING OF THE
     ACTIVITIES HAS BEEN CHANGED"
1050 T5=1
1060 FOR K= 2 TO E1
1070 Q(K)=1
1080 NEXT K
1090 FOR K=1 TO N1
1100 J2 = J(K)
1110 Q(J2)=0
1120 NEXT K
1130 FOR K= 2 TO E1
1140 IF Q(K)=0 THEN 1170
1150 PRINT "NO ACTIVITIES TERMINATE AT
     NODE ";K
1160 T5=0
1170 NEXT K
1180 E2=E1-1
1190 FOR K=1 TO E2
1200 Q(K)=1
1210 NEXT K
1220 FOR K=1 TO N1
1230 I2=I(K)
1240 Q(I2)=0
1250 NEXT K
1260 FOR K=1 TO E2
1270 IF Q(K)=0 THEN 1300
1280 PRINT "NO ACTIVITIES START AT NODE ";K
1290 T5=0
1300 NEXT K
1310 IF T5>0 THEN 1340
1320 PRINT "NETWORK IS INCOMPLETE"
```

```
1330 STOP
1340 REM                    ******
1350 REM                    STEP 0
1360 REM                    ******
1370 FOR L=1 TO E1
1380 R(L)=0
1390 S(L)=0
1400 T(L)=T9
1410 NEXT L
1420 K=1
1430 REM                    ******
1440 REM                    STEP 1
1450 REM                    ******
1460 J2=J(K)
1470 I2=I(K)
1480 S0=S(J2)
1490 IF S0>S(I2)+N(K) THEN 1550
1500 IF S0=S(I2)+N(K) THEN 1540
1510 S(J2)=S(I2)+N(K)
1520 R(J2)=R(I2)+V(K)
1530 GOTO 1550
1540 IF R(J2)<R(I2)+V(K) THEN 1520
1550 K=K+1
1560 REM                    ******
1570 REM                    STEP 2
1580 REM                    ******
1590 IF K<=N1 THEN 1460
1600 K=N1
1610 T(E1)=S(E1)
1620 REM                    ******
1630 REM                    STEP 3
1640 REM                    ******
1650 I2=I(K)
1660 J2=J(K)
1670 IF T(I2)<T(J2)-N(K) THEN 1690
1680 T(I2)=T(J2)-N(K)
1690 K=K-1
1700 REM                    ******
1710 REM                    STEP 4
1720 REM                    ******
1730 IF K>0 THEN 1650
1740 FOR K=1 TO N1
1750 I2=I(K)
1760 J2=J(K)
1770 C(K)=S(I2)
1780 D(K)=S(I2)+N(K)
1790 F(K)=T(J2)
1800 E(K)=T(J2)-N(K)
1810 H(K)=F(K)-C(K)-N(K)
1820 O(K)=MAX(0,0,0,S(J2)-T(I2)-N(K))
1830 P(K)=T(J2)-T(I2)-N(K)
1840 G(K)=S(J2)-S(I2)-N(K)
1850 NEXT K
1860 PRINT "RESULTS OF PERT"
1870 PRINT
1880 PRINT"ACTIVITIES"
1890 PRINT"  NO   START FINSH   DURATION
     OPTIMC  PESSMC MOST LY"
1900 PRINT"         EVENT EVENT     (MEAN)
     TIME      TIME    TIME"
1910 FOR K=1 TO N1
1920 PRINT K;I(K);J(K);N(K);A(K);B(K);M(K)
1930 NEXT K
1940 PRINT "ACTIVITY START AND FINISH TIMES"
1950 PRINT "  NO EARLY    EARLY    LATE
     LATE"
```

```
1960 PRINT"      START  FINISH    START
     FINISH"
1970 FOR K=1 TO N1
1980 PRINT K;C(K);D(K);E(K);F(K)
1990 NEXT K
2000 PRINT"FLOATS"
2010 PRINT"NO           TOTAL      FREE
     INDEPT    SAFETY"
2020 PRINT"           FLOAT      FLOAT
     FLOAT    FLOAT"
2030 FOR K=1 TO N1
2040 PRINT K;H(K);G(K);O(K);P(K);
2050 IF H(K)>0.001 THEN 2070
2060 PRINT"   C";
2070 PRINT
2080 NEXT K
2090 PRINT "CRITICAL PATH MARKED WITH C"
2100 PRINT "EVENT TIMES"
2110 PRINT "NO    EARLY   LATE   VARIANCE"
2120 PRINT "      START   START"
2130 FOR K=1 TO E1
2140 PRINT K;S(K);T(K);R(K)
2150 NEXT K
2160 PRINT "THE PROJECT HAS EXPECTED
     DURATION";S(E1)
2170 PRINT "THIS TIME HAS VARIANCE";R(E1)
2180 T5=SQR(R(E1))
2190 PRINT "AND STANDARD DEVIATION";T5
2200 STOP
2210 PRINT "TOO MANY EVENTS;  TRY AGAIN"
2220 T5=1
2230 RETURN
2240 PRINT "TOO MANY ACTIVITIES;  TRY AGAIN"
2250 T5=1
2260 RETURN
2270 PRINT "TOO FEW EVENTS;  TRY AGAIN"
2280 T5=1
2290 RETURN
2300 PRINT "TOO FEW ACTIVITIES;  TRY AGAIN"
2310 T5=1
2320 RETURN
2330 PRINT "TOO MANY EVENTS/TOO FEW
     ACTIVITIES;  TRY AGAIN"
2340 T5=1
2350 RETURN
2360 PRINT "START MUST BE LESS THAN FINISH"
2370 T5=1
2380 RETURN
2390 PRINT "ALL TIMES MUST BE POSITIVE
     OR ZERO"
2400 T5=1
2410 RETURN
2420 PRINT "TIMES MUST BE ORDERED A<=M<=B"
2430 T5=1
2440 RETURN
```

5.2.3 Limitations of PERT

PERT, like Critical Path Analysis, has become well established in commercial usage. However, it is not wholly without defects. Foremost of these is the assumption that the critical path that is found from the analysis is the path which will actually determine the duration of the project. This need not be the case, as the trivial network below shows. (Fig. 5.30).

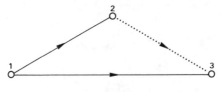

Fig. 5.30.

i	j	a	b	m	mean	variance
1	2	4	10	8	7.666	1
1	3	8	8	8	8	0
2	3	0	0	0	0	0

In this network, the critical path is the single activity (1,3). Using this, the expected duration of the project is 8 time units, with zero variance. However, the variance of the alternative path, (1,2), (2,3), is 1.0 with expected duration 7.666 time units. If a normal approximation is assumed for the duration of this path, then with probability 0.37 the project takes more than 8 time units.

The same effect may be found for larger networks and longer critical paths. Although it can cause problems in a few cases, it is not often a source of confusion. In most projects, the critical path dominates all other paths. Thus it is unlikely that any other set of activities will take longer than the activities on the critical path. In the rare cases where several paths have approximately the same expected length, it is possible to find an approximate measure of the distribution of the project's duration by simulation. (In this, a series of random numbers are generated, one for each activity in the project, to represent its duration. Then the critical path is found for the network with these durations. The process can be repeated several times, to give a number of possible project durations, as a random sample from the distribution of all possible durations.)

5.3 RESOURCE ALLOCATION AND SMOOTHING

In certain projects, the duration of activities can be reduced, by the expenditure of more money. This may mean that extra resources are diverted to some activity, or that overtime is worked on it. This extra expenditure will be of value if the project has to be completed within a particular time span, and the critical path without any acceleration takes longer than this. Alternatively, extra expense may be necessary if a part of a project has been delayed by some unforeseen event, and it is necessary to bring the whole project back to schedule.

When this possibility exists, it is usually assumed that the excess cost of the activity is a linear function of time, decreasing as the duration of the activity increases. So, for every unit reduction in the time of an activity, a constant extra cost, the **cost coefficient** for that activity, is incurred. It is further assumed that the time taken to complete an activity cannot be reduced or extended indefinitely. It must lie between a lower bound known as the **crash time** and an upper bound of the **normal time.** Then the objective of the analysis is to complete the project within the given limited time span, at minimum cost (which is effectively the same as saying at minimum extra cost compared with taking the normal time for each activity.)

This problem may be formulated as a linear program. The objective function is the cost, to be minimised, and the variables are the times taken for each activity, and the early start times for the events. These are subject to a set of linear constraints, of the forms:

crash time for activity (i,j) $\leqslant$ time for activity (i,j)
normal time for activity (i,j) $\geqslant$ time for activity (i,j)
early start for event j $\geqslant$ early start for event i + time for activity (i,j)

together with the constraint on the project duration:
(early start for final − early start for first event)
$\leqslant$ maximum time for the project.

However, such a formulation, though viable, is likely to be too cumbersome for most network problems. A more efficient alternative exists. The network is analysed using normal times for all activities. In this, all the activities on the critical path are listed, and that with the smallest cost coefficient is identified. This is selected as the first activity to be accelerated. It is evident that only the activities on the critical path need to be examined, since a reduction in the duration of any others would have no effect on the project duration, and would increase cost. It is also evident that where there is a choice, the activity with the smallest cost coefficient must be chosen. The duration of this activity is then reduced until:

(a)   the project's time constraint is satisfied
or (b)   the activity's crash time is reached
or (c)   (an)other path(s) become(s) critical
If (a),   then the problem has been solved.
If (b),   then the activity with the next smallest cost coefficient is selected, and the process repeated.
If (c),   then all the critical paths are considered together.

Ways of reducing the project duration will be: either reduce the duration of some shared activity, or reduce the duration of two activities, one on each critical path, or both of these. Each possibility is examined, and the one with the least total of cost coefficients is selected. Then the corresponding activity(ies) is (are) reduced, until one of the three possible reasons for stopping the reduction is encountered; then the process is repeated. Eventually condition (a) will be satisfied, assuming that there is a solution to the problem.

5.3.1 Worked Example

For the network shown below, (Fig. 5.31) the crash times, normal times and cost coefficients are as given in the table. How can the project be completed in 8 days at minimum cost?

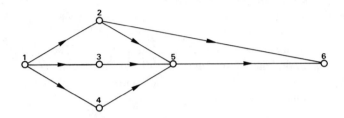

Fig. 5.31.

Activity		crash time	normal time	cost coefficient
i	j	(days)	(days)	(£100s)
1	2	1	2	4
1	3	2	6	5
1	4	2	3	2
2	6	1	3	5
2	5	2	4	1
3	5	1	2	10
4	5	1	2	8
5	6	2	3	3

The critical path is (1,3), (3,5), (5,6) with a duration of 11 days. The corresponding cost coefficients are 5, 10, 3 respectively, so the last one is selected. When the duration has been reduced from 3 to 2, the crash time is reached, and activity (1,3) is next selected. This may be reduced from 6 days to 4 days without any other path becoming critical, at which point the project duration will have been reduced to the required 8 days.

EXERCISES

(5.1)   Draw networks for the following projects, and label the events by the numbering algorithm.

     (a)   A student collects bread, butter, instant coffee and marmalade from a cupboard. He fills and boils a kettle. He finds a mug, plate, knife and spoon. He measures the instant coffee into the mug with the spoon, and adds boiling water, stirring with the spoon. He toasts the bread and spreads butter and marmalade on it. Then he enjoys his breakfast!

(b) A car-owner decides to change his car. He calculates how much he can afford, and makes a list of the kinds of cars to consider. He searches in the car-sales column of the newspaper and (eventually) decides to buy one. The seller arranges to check the car mechanically, and buys a new tax disc. The buyer pays a deposit, arranges for insurance cover, and sells his old car. The balance of the purchase money is paid, and the purchaser drives off.

(5.2) Find the characteristic times for each activity in the following networks, and identify the critical path:

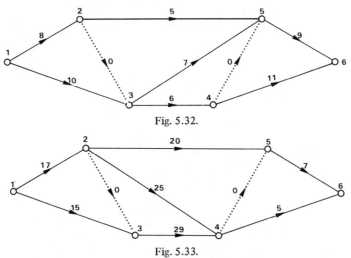

Fig. 5.32.

Fig. 5.33.

(5.3) What (if anything) is wrong with these networks?

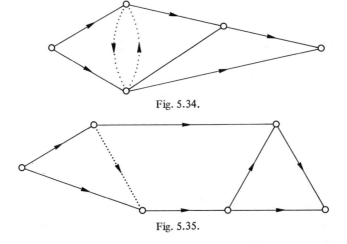

Fig. 5.34.

Fig. 5.35.

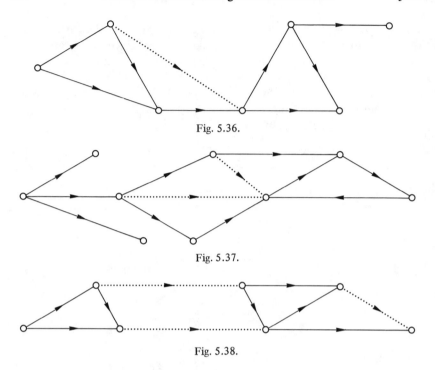

Fig. 5.36.

Fig. 5.37.

Fig. 5.38.

(5.4)   A project is being undertaken by two men who can either work together
on one activity or on two separate activities. Once an activity has been
started, the number working on it cannot be changed. The activities,
their predecessors, and the time taken (in man-days) are:

Activity	Preceded by	Duration
A	–	3
B	–	1
C	–	1
D	A,B	1
E	A,B	3
F	B	1
G	B,C	1
H	D	2
I	E,F	1
J	E,F,G	2
K	H,I	1
L	J	3

Draw the network for this project, and devise a schedule for the two
men to allow the work to be completed as quickly as possible.

(5.5)   In the network below, activities are subject to random delays, so that
the times a, b and m are as shown. Find the expected duration of the
project, and the probability that it is completed within:

(a)   37 days
(b)   39 days

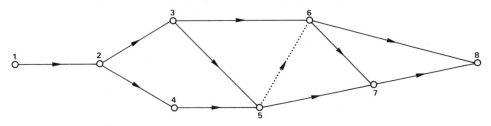

Fig. 5.39.

i	j	a	b	m
1	2	5	7	6
2	3	1	10	4
2	4	2	8	6
3	5	1	1	1
3	6	4	7	5
4	5	3	7	5
5	6	0	0	0
5	7	10	12	11
6	7	9	15	11
6	8	9	11	10
7	8	6	10	8

6

Tours in Networks

6.1 THE POSTMAN PROBLEM

In the course of delivering mail, a postman must gradually work along the length of each of a number of interconnected roads and streets. Mail will normally be delivered at several points along each of these. Where there are several streets meeting at a junction, the postman will have a choice of routes to follow. The 'Postman problem' is concerned with this choice. It may be stated as: which route should a postman follow in order that he should pass along each street at least once (so as to make deliveries), and return to his starting point, while travelling the least possible total distance?

This problem is not faced by postmen only; anyone who is planning a series of deliveries or collections along a number of streets is likely to be interested in the shortest tour which takes in all those streets. Refuse collection vehicles have to follow routes which pass along each street at least once; so do highway inspectors and (by broadening the idea of a street) inspectors of pipelines and power cables. Milk roundsmen and meter readers, traffic wardens and police patrols all face an essentially similar type of problem. However, the apparently similar problem faced by a road-sweeper is not quite the same; his problem is of sweeping both sides of each street, so requires to pass along each street at least twice. A tour with this requirement can be found much more easily than a postman tour.

In practice, there may be further constraints on the tour than those which are stated above; there may be a requirement that certain streets be visited in a particular order, or that returns to the starting point occur at specific times during the tour. These may mean that the shortest 'postman tour' cannot always be used, but it may be possible to modify the problem specification so that the algorithm can still be used.

This problem can be readily transformed into a network optimisation problem; streets can be transformed into arcs of a network, their junctions into nodes, and the distance travelled by the postman in making deliveries along a street into the length associated with that arc. So, for a network (N, A, D), the problem is that of identifying the circuit which traverses each arc at least once, and whose total distance is a minimum. The total length of such a circuit is evaluated by summing the lengths of the arcs multiplied by the number of times each is used. The circuit which results will be called a 'postman tour' for the network.

This problem is sometimes referred to as 'The Chinese Postman Problem', not because of any special behaviour when mail is delivered in China, but in recognition of the pioneer work reported in a Chinese mathematical journal (Kwan, M-K [27]). Several authors have since studied aspects of the problem, with varying requirements on the network, and a number of algorithms have been presented. (Edmonds & Johnson [24] present a good bibliography of this.) However, a much earlier study of a related problem gave rise to some of the results which are used in solving the postman problem. In 1736, the mathematician Leonhard Euler posed the 'Konigsberg bridge problem'. The town of Konigsberg (now Kaliningrad, in the USSR), is built on the banks of the river Pregel, with two islands which were linked to one another and the river banks by bridges, as shown in the plan. (Figs. 6.1 and 6.2). The problem posed by Euler was to find a way of crossing each bridge exactly once, and returning to the starting point. In terms of the postman problem, it is that of finding a postman tour of

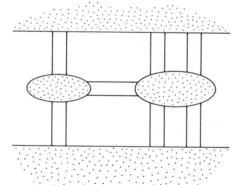

Fig. 6.1 – The seven bridges of Konigsberg.

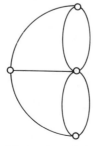

Fig. 6.2 – The graph equivalent to Fig. 6.1.

length seven units, in the network of undirected arcs shown in the figure, where each arc has unit length. Euler concluded (correctly) that no solution existed, and in the process laid some of the foundations of graph theory.

For a tour to exist in a network, the network must be connected; otherwise there can be no circuits and no postman tours. This property will be assumed in the ensuing consideration of the problem.

The simplest of all networks to deal with is one in which there is a postman tour which traverses each arc exactly once. The total length of such a tour will be the sum of all the lengths of the arcs in A, and it is evident that there can be no shorter tour. This special case is not affected by the lengths of the arcs; any, or all, may have their lengths altered, and the tour will be optimal, though of a different length. And so, this special case may be considered without the need to use the distance matrix D, using the graph (N,A) only. The tour which satisfies these simple constraints is known as an Euler (or Eulerian) tour, to mark Euler's contribution to the study of graphs. The graph shown in Fig. 6.3 possesses several such tours, while that in Fig. 6.4 has none.

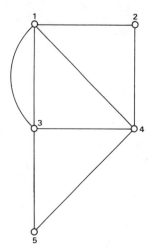
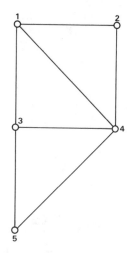

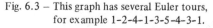

Fig. 6.3 – This graph has several Euler tours, for example 1-2-4-1-3-5-4-3-1.   Fig. 6.4 – This graph has no Euler tour.

For such networks, the problem of finding a postman tour is that of finding an Euler tour, and a straightforward algorithm exists for this. More difficult to deal with are those networks where there is no Euler tour, and some or all of the arcs are directed. The first case to be considered is that of finding a tour in an undirected graph, and the Euler tour is a special case of this.

First, it should be noted that the starting point of any postman tour does not affect the tour itself. Suppose that a tour has been found which starts and finishs at a given node, s, and a tour is required which starts and finishes at some other node, t. Then the tour from s must pass through t at some stage, and the first visit to t will divide the tour into two paths, P_1 and P_2. A postman tour originating at t can be formed by using the path P_2 followed by P_1.

6.1.1 Finding a postman tour in an undirected network

The first stage in finding tours in undirected networks is to determine whether the network is even or not; an even network is one in which the number of arcs which are incident on every node is even. If there are any nodes with an odd number of incident arcs, then the network is described as being not even (this term is used, rather than calling a network 'odd'!). In an even network, there will be Euler tours, and the algorithm will find one such, as the postman tour. In other networks, the algorithm will find the postman tour by finding those arcs which must be traversed more than once, and effectively creates an extended network with some duplicate arcs, and an Euler tour in this.

The first stage, that of finding whether the network is even or not, is simply a matter of counting the arcs which are incident on each node. If there is any node with an odd number, then no Euler tour exists. Nodes with an odd number of incident arcs occur in pairs. In the node-arc incidence matrix, there is an even number of non-zero entries, two for each arc, corresponding to the two end nodes of the arc. As each of the nodes with an odd number of 'incidences' accounts for an odd number of such entries, there must be an even number of such nodes in the graph.

For an even network, finding an Euler tour is straightforward. Arcs are divided into two sets, those which have been used and the remainder. A tour is built up by transferring arcs from the latter set to the former. Initially, all the arcs are in the second set. Starting with the origin of the desired tour, any unused arc incident on this node is selected. This arc becomes used, and the process is repeated, finding an unused arc incident on the last node to have been used. So the tour is built up, one arc at a time. Eventually the origin of the tour will be reached again; if there any unused arcs incident on this node, then the process continues, until the origin is reached, *and* all the arcs which are incident on it have been added to the tour. If, by this stage, all the arcs have been used, then the tour is complete. Otherwise, one or more extra parts must be 'spliced' into the tour; these are found by selecting a node on the tour which possesses an unused incident arc. This is used as the starting point for a tour of unused arcs, and this 'mini-tour' is inserted into the tour at the point where the original tour visited the selected node. The process of testing and generating continues until all the arcs have been used. Formally, this may be written as an algorithm:

Postman tour in an even, undirected network, (An Euler tour)

step 0   Let s be the origin of the tour. Make all arcs 'unused'. Let $t = s$ (t represents the last node visited). Let U and V be two empty sets of arcs, representing the partially complete tour and successive 'mini-tours' respectively.

step 1   Find any arc between t and q (another node) which is unused. Make it used, and add it to U. Set $t = q$.

step 2   If t equals s, do step 3; otherwise repeat step 1.

step 3   Insert U in V, at the point in V where node s is first reached; U becomes empty. Find a node t which is visited in V, but has unused arcs incident on it. If there is no such node, then stop; otherwise, set $s = t$, and return to step 1.

An alternative approach to this algorithm is to amend step 1 so that arc (t,q) is any unused arc which will satisfy the rule 'Do not select an arc which will make the unused part of the network unconnected'. This removes the need for step 3, since there will never be circumstances in which splicing of tours is necessary, while introducing a more detailed check at step 1. When finding tours by hand, it is a very easy rule to apply, since the used arcs can be erased from the network diagram, and the connectedness of the remaining arcs checked by eye!

When the network is not even, then some of the arcs will have to be repeated. The objective for the postman will be to select a set which has minimum total distance, since the length of the postman tour will be the total length of the arcs which are repeated, plus the fixed total length of all the arcs in the network. In order to find the best such set, attention is first given to the nodes of odd order. Since each visit to a node, including these, requires the use of two arcs, it is apparent that at least one of the arcs terminating at an odd-order node must be used twice. So, the first part of the method is given to finding all the odd-order nodes, and then to finding the shortest path distances between them. For, in order to complete a postman tour, the tour must include a number of paths between the odd-order nodes. These nodes are then matched in pairs, and the matching for which the total distance is least found. (If there are only two odd nodes, then there is only one possible matching; if there are four odd nodes, (a,b,c,d) say, there are three possible matchings $((a,b), (c,d); (a,c), (b,d); (a,d), (b,c))$; if there are six odd nodes, there are fifteen possible pairings; and, in general, if there are $2m$ odd nodes, there are $(2m)!/(m!2^m)$ possible matchings.) Once this optimal matching has been found, the paths which correspond to it are added to the original network, which then becomes an even network, and an Euler tour is found for this extended network.

The problem of finding the best possible matching of nodes is a special case of a range of more general problems, those of matching nodes or arcs to satisfy particular given requirements. Algorithms for solving such problems have been produced, and are detailed by, for example, Balinski [22]. For small and medium sized problems, complete enumeration is a quick, and not particularly inefficient, method for the special case of matching that is posed within the postman problem. The algorithm for the solution of the postman problem in any undirected network may now be given:

Postman tour in any undirected network:

step 0   Determine whether the order of each node i of the network (N,A,D) is odd or even. Let $S = (i_1, i_2, \ldots i_{2p})$ be the set of all odd-order nodes. Create the network $(N,A^*,D) = (N,A,D)$. If S is empty, go to step 3.

step 1   Using the matrix D of arc lengths, calculate the $2p * 2p$ matrix of shortest distances between members of S, using a shortest path routine.

step 2   Find the pairing of members of S which has minimal total length. Using this pairing, find the path which corresponds to these shortest distances, and add the arcs of this path to A^*.

step 3   Find an Euler tour in (N,A^*,D), which will be an even, undirected network.

6.1.2 Worked Examples

(Since the algorithm for the Euler tour is used as one step of the more general algorithm, the worked examples will use the latter.)

Example 1

For the network shown in Fig. 6.5.

$$N = \{1,2,3,4,5\}$$

$$A = \{(1,2), (1,3), (2,3), (2,4), (2,5), (3,4), (3,5)\}$$

$$D = \begin{bmatrix} 0 & 4 & 2 & \infty & \infty \\ 4 & 0 & 5 & 3 & 6 \\ 2 & 5 & 0 & 3 & 7 \\ \infty & 3 & 3 & 0 & \infty \\ \infty & 6 & 7 & \infty & 0 \end{bmatrix}$$

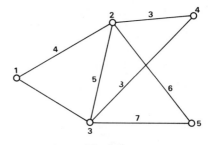

Fig. 6.5.

step 0   node 1 has order 2
node 2 has order 4
node 3 has order 4
node 4 has order 2
node 5 has order 2

$$S = \emptyset$$

$$(N,A^*,D) = (N,A,D)$$

S is empty, so go to step 3.

step 3   Use the algorithm for the Euler tour of $(N,A*,D)$.
step E0   Let $s = 1$, and let $t = s$. Let $U = V = \emptyset$. All of $A*$ becomes 'unused'.
step E1   Select arc $(1,2)$ and make it used. U becomes $\{(1,2)\}$ t becomes 2.
step E2   Repeat step 1.
step E1   Select arc $(2,5)$ and make it used. U becomes $\{(1,2),(2,5)\}$. t becomes 5.
step E2   Repeat step 1.
step E1   Select arc $(5,3)$ and make it used. U becomes $\{(1,2),(2,5),(5,3)\}$. t becomes 3.
step E2   Repeat step 1.
step E1   Select arc $(3,1)$ and make it used. U becomes $\{(1,2),(2,5),(5,3),(3,1)\}$. t becomes 1.
step E2   Since t equals s, do step 3.
step E3 V becomes $\{(1,2),(2,5),(5,3),(3,1)\}$.
          U becomes $\emptyset$. Both node 2 and node 3 have been visited in V, but retain unused incident arcs. Let $t = s = 2$, and do step 1.
step E1   Select arc $(2,4)$ and make it used. U becomes $\{(2,4)\}$. t becomes 4.
step E2   Repeat step 1.
step E1   Select arc $(4,3)$ and make it used. U becomes $\{(2,4),(4,3)\}$. t becomes 3.
step E2   Repeat step 1.
step E1   Select arc $(3,2)$ and make it used. U becomes $\{(2,4),(4,3),(3,2)\}$. t becomes 2.
step E2   Since t equals s, do step 3.
step E3 V becomes $\{(1,2),(2,4),(4,3),(3,2),(2,5),(5,3),(3,1)\}$, U becomes $\emptyset$. There are no nodes with incident, unused arcs, so V represents the postman tour in this network.

In this example, the nodes were all even, so there was no need to make use of the distance matrix D.

Example 2
For the network shown in Fig. 6.6.

$$N = \{1,2,3,4,5,6,7\}$$

$$A = \{(1,2), (1,3), (1,4), (1,5), (2,5), (2,7)$$
$$(3,4), (3,6), (4,5), (4,6), (5,7), (6,7)\}$$

$$D = \begin{bmatrix} 0 & 6 & 10 & 9 & 10 & \infty & \infty \\ 6 & 0 & \infty & \infty & 8 & \infty & 16 \\ 10 & \infty & 0 & 12 & \infty & 7 & \infty \\ 9 & \infty & 12 & 0 & 5 & 5 & \infty \\ 10 & 8 & \infty & 5 & 0 & \infty & 7 \\ \infty & \infty & 7 & 5 & \infty & 0 & 13 \\ \infty & 16 & \infty & \infty & 7 & 13 & 0 \end{bmatrix}$$

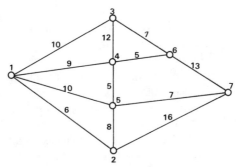

Fig. 6.6.

step 0   node 1 has order 4
          node 2 has order 3
          node 3 has order 3
          node 4 has order 4
          node 5 has order 4
          node 6 has order 3
          node 7 has order 3

$S = \{2,3,6,7\}$

$A^* = A$

step 1   Shortest distance matrix is

$$D^* = \begin{bmatrix} 0 & 16 & 18 & 15 \\ 16 & 0 & 7 & 20 \\ 18 & 7 & 0 & 13 \\ 15 & 20 & 13 & 0 \end{bmatrix}$$

step 2   The three possible pairings are:

(a) $(2,3)$ and $(6,7)$; length $= 16 + 13 = 29$

(b) $(2,6)$ and $(3,7)$; length $= 18 + 20 = 38$

(c) $(2,7)$ and $(3,6)$; length $= 15 + 7 = 22$

The best pairing is $(2,7)$ and $(3,6)$;
corresponding to this the paths are:

$(2,5)$ $(5,7)$ and $(3,6)$, so that $A^* = A + \{(2,5), (5,7), (3,6)\}$.

step 3   Find an Euler tour in (N, A^*, D).
          One such tour is:

$V = \{(1,2), (2,5), (5,7), (7,2), (2,5), (5,7), (7,6)$
$\quad\quad (6,3), (3,4), (4,6), (6,3), (3,1), (1,4), (4,5), (5,1)\}$

with a total length of 121 units.

6.1.3 Implementation

Most of the ideas which have been used in this algorithm can be translated into computer code with relative ease; for several, there is a choice of ways of constructing them. The tours are ordered sets, which can readily be stored as vectors of arc numbers, and a mini-tour can be spliced into another by moving part of the latter, and then copying the mini-tour into the gap which results. Alternatively, it could be stored by indexing each arc with its position in the tour; in PASCAL, the tour could be stored as an ordered series of records, each with a pointer to its successor. Storage of the set of arcs A^* could be by an extension of the set A, but in the PASCAL program presented below, a key has been introduced which initially records the number of times which each arc will be traversed. This is first set to one, and increased to two for each arc which has to be repeated as a result of step 2 of the algorithm. This key provides a mechanism for recording the used and unused arcs, since it is reduced by one for each traversal of an arc. The tour is completed when all arcs have key-value zero. The generation of all pairs of nodes is performed recursively in PASCAL; the set of nodes is scanned and the first member of the set is removed, along with each other member in turn. The set of nodes which is left is similarly scanned, until all nodes have been removed. The distance associated with this pairing is calculated, and compared with the smallest found so far. The best pairing is recorded in an array.

```
program prog6p1(input,output);
 const
        inf= 9999;
        maxarcs = 30;
        maxarcs2= 60;
        maxnodes = 30;
 type logarr = array[1..maxnodes] of boolean;
 intarr = array[1..maxnodes] of integer;
        dismat = array[1..maxnodes,
##    1..maxnodes] of integer;
        inset = 0..maxnodes;
        intset = set of inset;
        ar = array[1..maxarcs] of integer;
 ar2 = array[1..2,1..maxnodes] of inset;
 var lt,node2,node,t8,t9,len,minlen: integer;
        bigs,vset : intset;
        order,o : intarr;
        dis,dstar : dismat;
        i,j,lim,d : ar;
        arc,tour,extra,s,t,q,count,kount,noinv,
##    position,arcs,n : integer;
        nod,bestpair : ar2;
        t0 : inset;
 bigu,bigv : array[1..maxarcs2] of inset;

 function countset(n:integer; var
##    thisset:intset): integer;
    { countset takes as value the number of
##    members of set thisset }
    var i : inset;
        j : integer;
    begin
```

```
       j := 0;
       for i := 1 to n do if (i in thisset) then
##       j := j+1;
       countset := j
     end;
     function least(n:integer; var
##       thisset:intset): integer;
     { least takes the value of the numerically
##       least member of thisset }
     var j : inset;
     begin
       if (countset(n,thisset)>0) then
       begin
         j := 1;
         while (not(j in thisset)) do j := j+1;
         least := j
       end
       else least := 0
     end;

     function last(n:integer; thisset: intset) :
##       inset;
     { last is the largest member of thisset }
     var j: inset;
     begin
       if (countset(n,thisset)>0) then
       begin
         j:= n;
         while (not (j in thisset)) do j := j-1;
         last := j
       end else last := 0
     end { last } ;
     procedure generate(n: inset; var ns: inset;
##       var ss: intset; var nod: ar2);
     var i,j,k,l,jl,nss : inset;
         star,sstar : intset;
     {    generate is called recursively to
##       create and test all possible
     pairs of members of the set ss (which has n
##       members), and find the
     pairing which has minimal total distance }
     begin
       i := least(maxnodes,ss);
       l := (n-ns) div 2 +1;
       sstar := ss-[i];
       nod[1,l] := i;
       j := least(maxnodes,sstar);
       jl:= last(maxnodes,sstar);
       for i := j to jl do
       if (i in sstar) then
       begin
       nod[2,l] := i;
       star := sstar-[i];
       nss := ns-2;
if (nss>0) then generate(n,nss,star,nod)
       else
       begin
         len := 0;
         for k := 1 to (n div 2) do
len := len + dstar[nod[1,k],nod[2,k]];
         if (len < minlen) then
         begin
           minlen := len;
           for k := 1 to (n div 2) do
             for i := 1 to 2 do
               bestpair[i,k] := nod[i,k]
         end
```

```
      end
    end
 end { generate };

 procedure readnet;
 var t8 : integer;
 begin

    write('How many nodes, how many arcs?');
     readln(n,arcs);
       writeln('Enter arcs with their start
##    node finish node length');
        for t8 := 1 to arcs do
        repeat
           write('Arc number ',t8:3,'  ');
            readln(i[t8],j[t8],d[t8]);
    if ((i[t8]>n)or(j[t8]>n)) then
 writeln('Node number too high: try again');
    if ((i[t8]<0)or(j[t8]<0)) then
 writeln('Node number too low: try again');
    if (i[t8]=j[t8]) then
      writeln('start and finish must be
##    different: try again ')
     until ((i[t8]<>j[t8])and(i[t8]in[1..n])
            and(j[t8]in[1..n]));
 end {readnet} ;

 function min(a,b : integer):integer;
 begin
    if (a<b) then min := a else min := b
 end;
 procedure dijk(var o : intarr; var lt, t8:
##    integer ;
 n,s,t : integer; var d : dismat);
 var i,j,p,t2 : integer;
      q : logarr;
      l,r : intarr;
 begin


 for i := 1 to n do
 begin
   l[i] := inf;
   q[i] := false
 end;
 q[s] := true;
 l[s] := 0;
 p := s;


 repeat
 for i := 1 to n do
 if (not q[i]) then
 l[i] := min(l[i],l[p]+d[p,i]);
 p := 0;
 t2 := inf;
 for i := 1 to n do
 if ((not q[i])and(t2 >=l[i])) then
 begin
   p := i;
   t2 :=l[i]
 end;
 q[p] := true;


 until q[t];
```

```
for j := 1 to n do
if (q[j]) then
for i := 1 to n do
if ((i<>j)and(l[j] = l[i] + d[i,j])) then
##     r[j] := i;
t8 := n;
o[t8] := t;
i := r[t];
while (i<>s) do
begin
   t8 := t8-1;
   o[t8] := i;
   i := r[i]
end;
lt := l[t]
end; { of procedure dijk }

begin

{
Postman tour algorithm
      This finds an Euler tour in a graph,
##    which may have been extended
 to allow for repetitions of selected arcs,
 variables (all numerical variables are
##    integers)
 n      : number of nodes
 arcs   : number of arcs
 i,j,d  : vectors used to store the two ends
##    of the arc (i and j) and
             its length, in d
 lim    : vector used to record the number
##    of times that the corresponding
             arc should be traversed.   Usually
##    it will be one, but may be
             increased to allow the extension
##    of the graph for an optima tour
 bigs   : set of nodes with an odd number of
##    incidet arcs
 count  : number of members of bigs, used to
##    determine whether any
             extension is needed
 dis    : square matrix of distances
##    corresponding to i,j,d
 dstar  : square matrix with minimal
##    distances between members of bigs
 bestpair: 2 by (count div 2) matrix of
##    optimal pairing of nodes
 bigu   : ordered mini-tour of nodes
 bigv   : the postman tour, built up by
##    gradual assimilaion of mini-tours
 vset   : set of nodes so far visited in
##    bigv or the current bigu
 noinv  : number of arcs in v
 extra  : number of arcs to be repeated
 position: place in bigv into which the
##    current bigu will be spliced}

writeln('Postman tour in an undirected
##    network');
writeln;
readnet;
extra := 0;
bigs := [];
for node := 1 to n do order[node] := 0;
for arc := 1 to arcs do
begin
```

```
    order[i[arc]] := order[i[arc]] + 1;
    order[j[arc]] := order[j[arc]] + 1
    end;
    count := 0;
    for arc := 1 to arcs do lim[arc] := 1;
    for node := 1 to n do
      if (order[node] mod 2 = 1) then
      begin
        bigs := bigs + [node];
        count := count + 1
      end;
    if (count > 0) then begin
    {              ******
                   STEP 1
                   ******
    }
      for node := 1 to n do
      for node2 := 1 to n do
      begin
        dis[node,node2] := inf;
        dstar[node,node2] := inf
      end;
      for arc := 1 to arcs do
      begin
        dis[i[arc],j[arc]] := d[arc];
        dis[j[arc],i[arc]] := d[arc]
      end;
      for node := 1 to n do
      for node2 := node to n do
      begin
        if (node2>node) then
        dijk(o,lt,t8,n,node,node2,dis)
        else lt := 0;
        dstar[node,node2] := lt;
        dstar[node2,node] := lt
      end;
    {              ******
                   STEP 2
                   ******
    }

      minlen := inf;
      for node := 1 to n do
    for t0 := 1 to 2 do  bestpair[t0,node] := 0;
    t0 := count; generate(count,t0,bigs,nod);
    writeln('The optimal postman tour repeats
##      arcs with a total length',minlen);
      for node := 1 to count div 2 do
      begin
        dijk(o,lt,t8,n,bestpair[1,node],
##      bestpair[2,node],dis);
        o[t8-1] := bestpair[1,node];
        for t9 := t8 to n do
        begin
          arc := 1;
          while not
##    (((((o[t9-1]=i[arc])and(o[t9]=j[arc]))or
    ((o[t9]=i[arc])and(o[t9-1]=j[arc])))or
                   (arc>arcs)) do
          arc := arc+1;
    if (arc<=arcs) then lim[arc] := lim[arc]+1;
          extra := extra + 1
        end
      end;
    end;
    {              ******
                   STEP 3
                   ******
```

```
                         ************
                         EULER STEP 0
                         ************
       }
       for tour := 1 to maxarcs2 do
       begin
         bigu[tour] := 0;
         bigv[tour] := 0
       end;
       vset := [];
       s := 1;
       t := 1;
       Kount := 1;
       position := 0;
       noinv := 0;
       {                 ************
                         EULER STEP 1
                         ************
       }
       repeat
         repeat
           for arc := 1 to arcs do
             if ((lim[arc]>0)and((i[arc]=t)
                or(j[arc]=t))) then
             begin
               lim[arc] := lim[arc]-1;
               bigu[Kount] := arc;
               Kount := Kount +1;
               if (i[arc]=t) then q := j[arc];
               if (j[arc]=t) then q := i[arc];
               t := q;
               vset := vset + [q]
             end
       {                 ************
                         EULER STEP 2
                         ************
       }
         until (t=s);
       {                 ************
                         EULER STEP 3
                         ************
       }
         Kount := Kount-1;
       if (noinv>0) then
           for tour := noinv downto (position+1) do
             bigv[tour+Kount] := bigv[tour];
           for tour := 1 to Kount do
           begin
             bigv[position+tour] := bigu[tour];
             bigu[tour] := 0
           end;
         noinv := noinv + Kount;
         if (noinv < (arcs+extra)) then
         begin
       {
         find a node to start a mini-tour from
       }
           for arc := 1 to arcs do
             if ((lim[arc]>0)and((i[arc]in
##           vset)or(j[arc]in vset))) then
             begin
               if (i[arc] in vset) then t := i[arc];
               if (j[arc] in vset) then t := j[arc]
             end;
             s := t;
             Kount := 1;
```

```
position := -1; arc := 1; repeat
     if(((j[bigv[arc]]=t)or(i[bigv[arc]]=t))
##   and((j[bigv[arc+1]]=t)or
(i[bigv[arc+1]]=t))) then position := arc;
     if (t=1) then position := 0;
     { rather awkward test to find if the
##   two arcs meet at t }    arc := arc+1
     until ((position >=0)or(arc>noinv))
     end
until ((position < 0)or(noinv=(arcs+extra)));
   if (position < 0) then writeln('There is
##   no postman tour')
   else    begin
        writeln('A postman tour is');   t := 1;
        for arc := 1 to noinv do
        begin
           if (t=i[bigv[arc]]) then q :=
##        j[bigv[arc]] else q := i[bigv[arc]];
           writeln(t,q);    t := q    end;
     end
  end.
```

6.1.4 Postman tours in networks with directed arcs

When the arcs of the network (N,A,D) are not all undirected, then it is more difficult to find a postman tour. It is convenient to consider such networks in two groups; first, those which have all their arcs directed, and then those which are composed of a mixture of directed and undirected arcs. In contrast to the situation described earlier, there may not be a postman tour in such networks, since it is possible for a node, or several nodes, to be linked to the remainder of the network in such a way that all the arcs between these two parts are in one direction — the postman becomes trapped! This is demonstrated in the network in Fig. 6.7, in which the only arcs between nodes 1 and 2 and the remainder of the network lead away from the nodes; once the postman has left either node 1 or node 2 he can never return.

In a network whose arcs are all directed, two possible situations occur: if the number of arcs entering a node is equal to the number leaving it, for every node of the network, then there is an Euler tour; if not, there is not such a tour. The first of these situations is often referred to as being symmetric.

In a symmetric, directed network, the postman tour can be found using the same method as was described for the Euler tour in an undirected network. A tour can be built up by selecting a starting node, and leaving it by any unused arc whose sense is away from that node. The arc becomes used, and the process is repeated until the starting node is reached again. If necessary, mini-tours can be spliced into the tour, to use arcs which were unused. The chief difference between this rule for constructing a tour, and that explained earlier is that the sense of the tours becomes important; whereas a second Euler tour can be constructed by reversing the circuit when all arcs are undirected, this is not the case with a directed network.

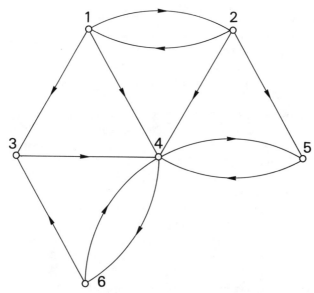

Fig. 6.7 – There is no postman tour since it is impossible to return to nodes 1 and 2.

When the network is not symmetric, some arcs need to be repeated by the postman. In contrast to the problem of repeating arcs in an undirected network, where the greatest number of times any arc was traversed is two, in a directed network an arc may be traversed any number of times. In Fig. 6.8 a network is shown in which arcs $(1,2)$ and $(2,3)$ are each used n times, and n can

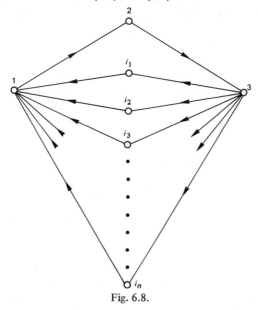

Fig. 6.8.

be increased without limit, by adding arcs and nodes which are similar to those indicated. In order to find a postman tour in a network, it is first necessary to calculate how many times each arc has to be used.

Once this is known, an Euler tour can be found in the symmetric network which is the result of extending the network (N, A, D) by the appropriate number of copies of each arc.

Such a network is known not to be symmetric because of an imbalance of arcs into and out of some of the nodes. For each node i in N, consider the two sets $S_{in}(i)$ and $S_{out}(i)$. $S_{in}(i)$ is the set of arcs which terminate at i (that is, $S_{in}(i) = \{(j,i) : (j,i) \text{ in } A\}$. In the same way, $S_{out}(i)$ is the set of arcs which originate at i ($S_{out}(i) = \{(i,j) : (i,j) \text{ in } A)\}$. If $S_{in}(i)$ is smaller than $S_{out}(i)$, then some of the members of $S_{in}(i)$ will have to be repeated, otherwise members of $S_{out}(i)$ will remain unused, and vice versa. Suppose there is associated with arc (j,k) a number $x(j,k) + 1$ which records the number of times that the arc is used (as in the computer program for the undirected network). Clearly $x(j,k) \geqslant 0$. The number of times that node i is visited is given either by the number of times it is reached by the postman, or by the number of times it is left by the postman. These two must be equal, and so: number of times node i is reached =

$$| S_{in}(i)| + \sum_j x(j,i) = | S_{out}(i)| + \sum_j x(i,j) \qquad (1)$$

= number of times the postman leaves node i

But $x(i,j)$ must also be chosen to minimise the sum (i,j) in $A \sum x(i,j) * d(i,j)$. This same objective function has been encountered earlier, as a minimum cost flow problem, with x representing the flow and d the cost. Rearranging (1) gives the imbalance of flow at node i, as $\sum_j (x(j,i) - x(i,j)) = | S_{out}(i)| - | S_{in}(i)| = Y(i)$.

For flow to be conserved in the minimum cost flow problem, node i must be regarded as a source ($Y(i) < 0$), or as a sink ($Y(i) > 0$), or as neither ($Y(i) = 0$). To convert the problem into one of a minimum cost feasible flow, those nodes, i, which act as sources must be linked to a super-source by arcs whose flow is fixed at $| Y(i)|$, and a similar set of arcs used to link sinks to a super-sink. Super-source and super-sink are linked in the usual way to maintain flow. Once the minimum cost flow problem has been solved for $x(i,j)$, this number of copies of arc (i,j) are added to the network, and an Euler tour found in the symmetric network of directed arcs which results.

The algorithm for a postman tour in any directed network may thus be stated formally:

step 0   For each node i in the network (N, A, D), calculate $Y(i)$ = number of arcs leaving i — number of arcs entering i. Create the network $(N, A^*, D) = (N, A, D)$. If $Y(i) = 0$ for all nodes i, go to step 2.

step 1   For each arc (i,j) in (N,A,D) create a lower bound $l(i,j) = 0$, an upper bound $u(i,j) = \infty$, and a cost $c(i,j) = d(i,j)$ ($=$ length of the arc).
Add a new node r (to be both super-source and super-sink). Add arcs as follows:

For all i with $Y(i) < 0$, create an arc (r,i) with $l(r,i) = u(r,i) = |Y(i)|$ and $c(r,i) = 0$.
For all i with $Y(i) > 0$, create an arc (i,r) with $l(i,r) = u(i,r) = + Y(i)$ and $c(i,r) = 0$.
Solve the minimal cost feasible flow problem in the graph (N,A) with the given costs and constraints, resulting in flows $x(i,j)$ in arc (i,j).
Add $x(i,j)$ copies of arc (i,j) to A^*.

step 2   (Find an Euler tour in the directed graph (N,A^*)). This step is identical to the algorithm for an Euler tour in an even, undirected network, except that in step 1 of that algorithm the next arc must be (t,q). It is not possible to use 'any arc between t and q'.

6.1.5 Worked Example
For the network of Fig. 6.9.

$$N = \{1,2,3,4,5,6\}$$

$$A = \{(1,2), (2,3), (2,6), (3,6), (4,3), (5,3), (5,4), (6,4)$$
$$(6,5), (6,1)\}$$

$$D = \begin{bmatrix} 0 & 10 & \infty & \infty & \infty & \infty \\ \infty & 0 & 14 & \infty & \infty & 12 \\ \infty & \infty & 0 & \infty & \infty & 11 \\ \infty & \infty & 5 & 0 & \infty & \infty \\ \infty & \infty & 6 & 15 & 0 & \infty \\ 8 & \infty & \infty & 13 & 10 & 0 \end{bmatrix}$$

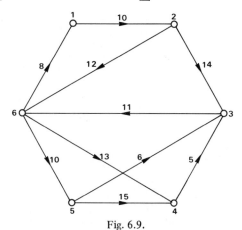

Fig. 6.9.

step 0 $Y(1) = 0$
$Y(2) = 1$
$Y(3) = -2$
$Y(4) = -1$
$Y(5) = 1$
$Y(6) = 1$

Create $(N, A^*, D) = (N, A, D)$.

step 1   Solve the minimum cost feasible flow problem for the network shown in Fig. 6.10.
The solution is: $x(i, j) = 0$ except for:

$x(1,2) = 1$; $x(3,6) = 3$; $x(4,5) = 1$; $x(6,5) = 1$; $x(6,1) = 1$
A^* becomes $A + \{(1,2), (3,6), (3,6), (3,6), (4,5), (6,5), (6,1)\}$.

step 2   Find an Euler tour in the network $(N, A, *, D)$ shown in Fig. 6.11. For instance, one such tour is:

$\underline{(1,2)}\ \underline{(2,3)}\ \underline{(3,6)}\ \underline{(6,5)}\ \underline{(5,3)}\ (3,6)\ (6,5)\ \underline{(5,4)}\ \underline{(4,3)}\ (3,6)$
$\underline{(6,4)}\ (4,3)\ (3,6)\ \underline{(6,1)}\ (1,2)\ \underline{(2,6)}\ (6,1)$

(the underlined arcs are those which are in the original network).

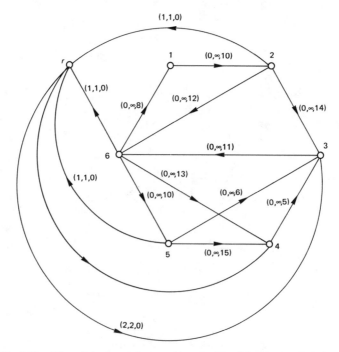

Fig. 6.10 – The minimal-cost-flow problem for Fig. 6.9. Arcs labelled (l, u, c).

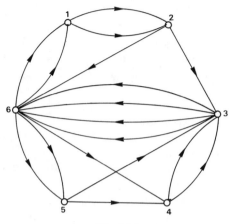

Fig. 6.11.

When the network contains a mixture of directed and undirected arcs, there is currently no algorithm which will find the postman tour in all possible cases. For certain special cases, there are algorithms, which are identified and described in such works as Edmonds and Johnson [24].

6.2 THE TRAVELLING SALESMAN PROBLEM

Unlike postmen, travelling salesmen usually concentrate their time on specific towns and locations, rather than the roads and streets which link them. The salesman wishes to visit his clients once, and to plan his route so as to minimise the total distance travelled. At each visit, he has a choice of which one to visit next. Yet, he still wishes to return to his starting point at the end of his travels. So, the 'travelling salesman problem' is that of choosing a route which visits each town (or client) once, and returns to the start, while covering the minimum total distance.

This problem is not confined to salesmen planning calls on clients; drivers of delivery vehicles who have to visit a number of customers in several localities may also wish to minimise their travelling distance, or their travel time. So may operators of vehicles, such as bulk milk lorries, which make a tour of collection points. In these practical areas, there are other constraints, but, nonetheless, the travelling salesman problem frequently forms a part of the algorithms which are used for finding the optimal route for the vehicles. The same problem also arises in a number of industrial problems. In some cases, it is desirable to plan a cyclic schedule of production on a single machine, making one product, changing to another, then another, and eventually returning to the first. The change-over times between different products may mean that some orders for the cycle are quicker than others, and the algorithm for the travelling salesman problem can be used for finding the best. In other applications, the best order for an automatic drill to drill a sheet of metal may be sought; the times taken to move the drill head between different sites will give the distance measure for the travelling salesman problem. It is even possible to use the formulation of this problem as a means of minimising the amount of wallpaper wasted when papering a wall with a patterned design!

6.2.1 Formulation as a network problem

In much the same way as for the postman problem, the travelling salesman problem can be readily transformed into a network problem. The towns to be visited may be identified with the nodes of a network; the possible routes between them as the arcs of that network; and the distances between towns represented as the arc lengths. The objective is to find a tour which passes through each node at least once and returns to its starting point, and which is shorter than every other such tour. In some cases, the tour (which is a circuit of the graph), is required to visit each town once and once only. Such a circuit is known as a **Hamiltonian circuit**, after the famous nineteenth century Irish mathematician, Sir William Hamilton. In most cases, the solution to a salesman problem is a Hamiltonian circuit. (The circumstances where this is otherwise correspond to this: for a town, j, which is visited twice, between i and k on the first visit, and between h and l on the second, the distances $d(i, k)$ and $d(h, l)$ must be greater than $(d(i,j) + d(j,k))$ and $(d(h,j) + d(j,l))$ respectively. The salesman, being no fool, knows that if it is quicker to go through a town rather than round it, then through it is the best way). It is possible to formulate the salesman problem as that of finding the shortest Hamiltonian circuit, if all the distances in D are replaced by the shortest path distance between the same nodes. For most purposes associated with men or vehicles travelling, the distance matrix will be symmetric, so that there are two equally short circuits, one the reverse of the other. This need not always be the case, and the matrix D may not be symmetric (as may happen where there are one-way streets, or, in the scheduling application, where the time to switch between two products depends on which is first and which second).

There have been a wide range of attempts to solve this problem, using several approaches from general optimisation techniques as well as from specifically network methods. Integer and dynamic programming can be used but both lead to very large programming problems, for even a small number of towns. (If there are n towns to be visited, then there will be $(n - 1)!$ possible tours. If the matrix D is symmetric, then this number may be halved, but both numbers are too large for every possible tour to be considered.) The methods that have been developed for finding solutions to the travelling salesman problem can be divided into two complementary groups. On one hand, there are several methods which will always find the optimal solution, but which may require a very large number of calculations to do so; for large problems, the time taken by such methods may render them of little value. On the other hand, there are methods which can find a good solution very quickly, but which may not find the best solution at all.

In each group, there are several methods available, and it is not possible here to consider them all. One representative method from each group will be used to demonstrate possible approaches to the problem.

6.2.2 Approximate methods

Within the family of approximate methods (those which give a solution, usually a good solution, fairly rapidly) there is yet another division. Some methods will find an exact solution, given enough time, while others can find a reasonable tour, and then terminate. The method of 'two-optimality' is in the latter category. In this method, a tour is created in some arbitrary way. Then two links are broken, and the paths that are left are joined up so as to form a new tour. If the length of this tour is less than the length of the original tour, then the new one is retained. Then two of its links are broken. The breaking and reassembly of links is carried out systematically, and eventually a tour is found which cannot be improved by the interchange of any two links. Such a tour is known as 'two-optimal'. The same approach can be used to find a 'three-optimal' tour, when three links are broken, and the tours which can be formed by reassembling the paths compared in length with the original tour's length. A tour which cannot be improved by any such rearrangement is three-optimal. (It is possible to continue in this way, but it is generally more efficient to use some other method; an optimal salesman tour through n towns is n-optimal, as it cannot be improved by any rearrangement of n links.)

There are several ways in which the initial tour can be established. The towns could be visited in alphabetical order, which is unlikely to be very efficient (for example, Birmingham, Brighton, Cardiff, Dover, Exeter); a simple rule of thumb is normally sufficient, such as the 'nearest-neighbour' method. In this, a starting point is chosen, and the tour is built up by visiting the nearest town which has not yet been visited until the start is reached. (One of the difficulties of this method is that the final link may be excessively long.)

So, a formal algorithm may be presented:

step 0   (Create an initial tour.) Select a starting point s. Choose t, so that $d(s,t) \leqslant d(s,j)$ for all $j \neq t$. Set $l = t$, and make nodes s and t 'visited'.

step 1   Select t from the unvisited nodes so that $d(l,t)$ is least. Add t to the end of the tour and set $l = t$. If there are further unvisited nodes, repeat step 1, otherwise add s to the tour and do 2.

step 2   The visited tour will be an ordered set of nodes:

$x_1, x_2, x_3, \ldots, x_n, x_1$, with total length L.

Set $i = 1$.

step 3   Set $j = i + 2$.

step 4   Consider the tour

$$x_1, x_2, \ldots x_i, x_j, x_{j-1}, \ldots x_{i+1}, x_{j+1}, x_{j+2}, \ldots x_1$$

created by exchanging the links (x_i, x_{i+1}) and (x_j, x_{j+1}). If this has length less than L, then make this the new tour, and do step 2.

step 5   Set $j = j + 1$. If $j \leqslant n$, do step 4. Otherwise, set $i = i + 1$. If $i \leqslant n - 2$, do step 3. Otherwise stop.

6.2.3 Worked Example of the two-optimal method

Suppose that a two-optimal tour is desired in the network whose distance matrix is: (Fig. 6.12).

$$\begin{bmatrix} \infty & 13 & 12 & 18 & 7 & 14 \\ 13 & \infty & 21 & 26 & 15 & 25 \\ 12 & 21 & \infty & 11 & 6 & 4 \\ 18 & 26 & 11 & \infty & 12 & 14 \\ 7 & 15 & 6 & 12 & \infty & 9 \\ 14 & 25 & 4 & 14 & 9 & \infty \end{bmatrix}$$

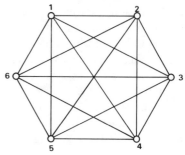

Fig. 6.12.

The initial tour is set up in steps 0 and 1:

step 0   Let $s = 1$. Then $t = 5$, and the partial tour is 1-5.
step 1   Set $t = 3$. The partial tour is now 1-5-3.
step 1   Set $t = 6$. The partial tour is 1-5-3-6.
step 1   Set $t = 4$. The partial tour is now 1-5-3-6-4.
step 1   Set $t = 2$. The partial tour is now 1-5-3-6-4-2.
          There are no unvisited nodes, so the tour is established as 1-5-3-6-4-2-1.
step 2 $i = 1$ $L = d(1,5) + d(5,3) + d(3,6) + d(6,4) + d(4,2) + d(2,1) = 70$
step 3 $j = 3$.
step 4   Consider the tour 1-3-5-6-4-2-1. This has length 80.
step 5 $j = 4$.
step 4   Consider the tour 1-6-3-5-4-2-1. This has length 77.
step 5 $j = 5$.
step 4   Consider the tour 1-4-6-3-5-2-1. This has length 70.
step 5 $j = 6$.
step 4   Consider the tour 1-2-4-6-3-5-1. This has length 70.
step 5 $j = 7$, set $i = 2$.
step 3 $j = 4$.
step 4   Consider the tour 1-5-6-3-4-2-1. This has length 70.
step 5 $j = 5$.
step 4   Consider the tour 1-5-4-6-3-2-1. This has length 71.

step 5 $j = 6$.
step 4   Consider the tour 1-5-2-4-6-3-1. This has length 78.
step 5 $j = 7$, set $i = 3$.
step 3 $j = 5$.
step 4   Consider the tour 1-5-3-4-6-2-1. This has length 76.
step 5 $j = 6$.
step 4   Consider the tour 1-5-3-2-4-6-1. This has length 86.
step 5 $j = 7$, set $i = 4$.
step 3 $j = 6$.
step 4   Consider the tour 1-5-3-6-2-4-1. This has length 86.
step 5 $j = 7$, set $i = 5$, stop.

In this case the original tour turns out to be two-optimal although there are other tours which are equally short.

6.2.4 Notes

When the distance matrix is symmetric, it is not necessary actually to measure the length of the tour in step 4. Instead, it would be sufficient to measure the change in the tour length by calculating:

$$d(x_i, x_{i+1}) + d(x_j, x_{j+1}) - d(x_i, x_j) - d(x_{i+1}, x_{j+1})$$

If this is positive, then the tour

$$x_1, x_2, \ldots x_i, x_j, x_{j-1}, \ldots x_{i+1}, x_{j+1}, x_{j+2}, \ldots x_1$$

should be stored as the best tour so far found, and the algorithm continued from step 2.

The algorithm generates the mirror image of the original tour when $i = 1$ and $j = n$. For a symmetric matrix, this need not be measured, since its length will be identical with the original length L.

6.2.5 Implementation of the two-optimal method

The two-optimal algorithm is made up of a series of nested loops, with i and j varying. These are very easily converted into loops in the program. Within these, it is necessary to have a means of generating test tours from the currently stored optimum, and of calculating the length of these. Neither of these should be difficult. A tour corresponds to an ordered set of node numbers, and in both PASCAL and BASIC, such ordered sets are most easily stored as vectors. Generating a new tour from an old one is simply a matter of directly copying some of the old one to the new, and of copying the central part in reverse order. So, in the code corresponding to step 4, there are three loops for copying the tour vectors; two copy the ends which are unchanged, while the third exchanges the appropriate nodes. As an added refinement, the PASCAL program has a defined variable TYPE for the tours, so that a whole tour can be copied without the need for a loop.

In both languages, the programs have been set up for symmetric matrices, and the distance matrix is modified using Floyd's algorithm (Chapter 3) so that it corresponds to the shortest distance matrix for all pairs of towns. The code for these processes can be removed if either non-symmetric matrices are used, or a Hamiltonian path is desired.

```
program prog6p2(input,output);
 const maxnodes = 30;
       mxnodesp= 31;
       maxarcs  = 30;
       inf= 9999;
 type   dismat = array[1..maxnodes,
##      1..maxnodes] of integer;
 tours  = array[1..mxnodesp] of integer;
 ar     = array[1..maxarcs] of integer;
        inset  = 0..maxnodes;
        intset = set of inset;
 var    tour,newtour : tours;
        dis : dismat;
        i,j,d : ar;
        bigs : intset;
        arc,arcs,n,t0,t1,t2,t,q,pos,minlen,
##      tourl,newtourl : integer;
        nochange : boolean;

 procedure readnet;
 var t8 : integer;
 begin

 repeat
    write('How many nodes, how many arcs?');
       readln(n,arcs);
    if (arcs>maxarcs) then
       writeln('Too many arcs:   change the
##     constant maxarcs');
    if (n>maxnodes) then
       writeln('Too many nodes:  change the
##     constant maxnodes');
    if (arcs<1) then
       writeln('Number of arcs must be
##     positive:  try again');
    if (n<1) then
       writeln('Number of nodes must be
##     positive:  try again')
    until ((arcs in [1..maxarcs])and(n in
##     [1..maxnodes]));
    writeln('Enter arcs with their start node
##     finish node length');
    for t8 := 1 to arcs do
    repeat
       write('Arc number ',t8:3,'  ');
       readln(i[t8],j[t8],d[t8]);
       if ((i[t8]>n)or(j[t8]>n)) then
 writeln('Node number too high: try again');
       if ((i[t8]<=0)or(j[t8]<=0)) then
 writeln('Node number too low: try again');
       if (i[t8]=j[t8]) then
          writeln('start and finish must be
##     different: try again ')
    until ((i[t8]<>j[t8])
    and(i[t8]in[1..n])and(j[t8]in[1..n]))
    end {readnet} ;
```

```
       function length(var thistour : tours; n :
##       integer) : integer :
       var l,t : integer;
       begin
          l := 0;
          for t := 1 to n do l := l +
##         dis[thistour[t],thistour[t+1]];
          length := l
       end ( length );
       begin
       (Two-optimal method for the traveling
##       salesman problem
       This program reads in the arcs with their
##       distances, and then
       calculates the shortest distance matrix for
##       the nodes, using
       Floyd's method.   It then finds the
##       two-optimal tour for this
       matrix.
       Variables used:
       n       : the number of nodes
       arcs    : the number of arcs
       i,j     : vectors to identify the ends of
##       the arcs
       d       : vector to hold the length of the arc
                 variables used in the initial
##       (nearest-neighbour) tour
       t       : the last node of a partially
##       completed tour
       q       : the node about to be added to the
##       tour
       pos     : the position (rank) of q in the tour
       bigs    : set of nodes which are not in the
##       partially completed tour
       vaiables used in the two-optimal method
       tour    : vector of length n+1 to hold the
##       best tour found so far
       newtour: similar vector for the tour being
##       examined
       tourl   : length of the tour stored in tour
       newtourl : length of the tour stored in
##       newtour
       nochange : logical variable to identify
##       when the tour is two-optimal
                  nochange is true if a pass
##       through all permitted changes
       to tour fails to find a shorter tour
       other variables for loop counters etc.
       }
       writeln('Two-optimal method for traveling
##       salesman problem');
       readnet;
       for t0 := 1 to n do
       for t1 := 1 to n do
          dis[t0,t1] := inf;
       for arc := 1 to arcs do
       begin
          dis[i[arc],j[arc]] := d[arc];
          dis[j[arc],i[arc]] := d[arc]
       end;
       for t0 := 1 to n do
       for t1 := 1 to n do
       for t2 := 1 to n do
       if (dis[t1,t2]>(dis[t1,t0]+dis[t0,t2])) then
              dis[t1,t2] := dis[t1,t0]+dis[t0,t2];
       ( simplified form of Floyd method
```

```
                ******
                STEP 0
                ******
   }
   t := 1;
   pos := 1;
   tour[1] := 1;
   bigs := [] ; for t0 := 2 to n do bigs :=
##    bigs +[t0];
   {
                ******
                STEP 1
                ******
   }
   repeat
     pos := pos+1;
     minlen := inf;
     for t0 := 1 to n do
     if ((t0 in bigs) and (dis[t,t0]<=minlen))
##     then
     begin
       minlen := dis[t,t0];
       a := t0
     end;
     tour[pos] := a;
     t := a;
     bigs := bigs - [a]
   until (pos=n);
   tour[n+1] := 1;
   {
                ******
                STEP 2
                ******
   }
   tourl := length(tour,n);
   repeat
   nochange := true;
     for t0 := 1 to n-2 do
     begin
   {
                ******
                STEP 3
                ******
   }
       for t1 := t0+2 to n do
       begin
   for t2 := 1 to t0 do newtour[t2] := tour[t2];
         for t2 := t1+1 to n+1 do newtour[t2]
##     := tour[t2];
         for t2 := t0+1 to t1 do newtour[t2]
##     := tour[t1+t0+1-t2];
   {
                ******
                STEP 4
                ******
   }
       newtourl := length(newtour,n);
       if (newtourl<tourl) then
       begin
         tourl := newtourl;
   for t2 := 2 to n do tour[t2] := newtour[t2];
         nochange := false
       end
     end
   end
   until (nochange);
   {
```

```
      ******
      STEP 5
      ******
   }
   writeln('Tour found');
   for t0 := 1 to n do
   begin
      write(tour[t0]:3);
      if (dis[tour[t0],tour[t0+1]])>=inf) then
         write('forced link')
   end;
   writeln(tour[n+1]:3);
   writeln('This has length ',tourl)
   end.
```

```
90 REM PROG6P3
100 REM TWO-OPTIMAL METHOD FOR
    TRAVELLING SALESMAN PROBLEM
110 REM
120 REM VARIABLES USED
130 REM
140 REM I,J,D : VECTORS TO HOLD DETAILS
    OF THE NETWORK ON
150 REM         INPUT
160 REM E      : DISTANCE MATRIX
170 REM B      : VECTOR USED FOR BUILDING
    UP AN INITIAL TOUR
180 REM          B(I)=0 IF I IS IN THE
    TOUR, 1 OTHERWISE
190 REM T      : VECTOR TO HOLD THE BEST
    TOUR FOUND SO FAR
200 REM U      : VECTOR TO HOLD THE
    CURRENT ALTERNATIVE TOUR
210 REM T HAS LENGTH L
220 REM U HAS LENGTH L1
230 REM N      : NUMBER OF NODES
240 REM N9     : SET TO 1 IF THE TOUR IS
    CHANGED
250 REM T0,...,N0,... WORK SPACE
260 DIM I(35),J(35),D(35),E(30,30),
    B(30),T(31),U(31)
270 I9=9999
280 PRINT "TWO-OPTIMAL METHOD FOR THE
    TRAVELLING SALESMAN PROBLEM"
290 PRINT"HOW MANY NODES, HOW MANY ARCS?";
300 INPUT N,A
310 IF N>30 THEN 1610
320 IF N<1 THEN 1630
330 IF A>35 THEN 1650
340 IF A<1 THEN 1670
350 N1 = N+1
360 PRINT"ENTER ARCS WITH START NODE,
    FINISH NODE, LENGTH"
370 FOR A1 = 1 TO A
380 PRINT "ARC NUMBER ";A1;" ";
390 INPUT I(A1),J(A1),D(A1)
400 IF I(A1)>N THEN 460
410 IF I(A1)<1 THEN 480
420 IF J(A1)>N THEN 460
430 IF J(A1)<1 THEN 480
440 IF I(A1)=J(A1) THEN 500
450 GOTO 520
460 PRINT "NODE NUMBER TOO HIGH : TRY AGAIN"
```

```
470 GOTO 380
480 PRINT "NODE NUMBER TOO LOW.   TRY AGAIN"
490 GOTO 380
500 PRINT "START AND FINISH MUST BE
    DIFFERENT.   TRY AGAIN"
510 GOTO 380
520 NEXT A1
530 FOR T0 = 1 TO N
540 FOR T1 = 1 TO N
550 E(T0,T1)=I9
560 NEXT T1
570 NEXT T0
580 FOR A1= 1 TO A
590 I1=I(A1)
600 J1=J(A1)
610 E(I1,J1)=D(A1)
620 E(J1,I1)=D(A1)
630 NEXT A1
640 FOR T0 = 1 TO N
650 FOR T1 = 1 TO N
660 FOR T2 = 1 TO N
670 T3=E(T1,T0)+E(T0,T2)
680 IF E(T1,T2)<T3 THEN 700
690 E(T1,T2)=T3
700 NEXT T2
710 NEXT T1
720 NEXT T0
730 REM               ******
740 REM               STEP 0
750 REM               ******
760 FOR T0 = 1 TO N
770 E(T0,T0) = I9
780 NEXT T0
790 T3=1
800 P=1
810 T(1)=1
820 B(1)=0
830 REM               ******
840 REM               STEP 1
850 REM               ******
860 FOR T0=2 TO N
870 B(T0)=1
880 NEXT T0
890 P=P+1
900 M=I9
910 FOR T0 = 1 TO N
920 IF B(T0)=0 THEN 960
930 IF E(T3,T0)>M THEN 960
940 M=E(T3,T0)
950 Q=T0
960 NEXT T0
970 T(P)=Q
980 T3=Q
990 B(Q)=0
1000 IF P<N THEN 890
1010 T(N1)=1
1020 L=0
1030 FOR T0=1 TO N
1040 T1=T0+1
1050 T2=T(T0)
1060 T3=T(T1)
1070 L=L+E(T2,T3)
1080 NEXT T0
1090 REM               ******
1100 REM               STEP 2
1110 REM               ******
1120 N2=N-2
```

```
1130 N9=0
1140 FOR T0=1 TO N2
1150 REM                   ******
1160 REM                   STEP 3
1170 REM                   ******
1180 T3=T0+2
1190 FOR T1=T3 TO N
1200 FOR T2 = 1 TO T0
1210 U(T2)=T(T2)
1220 NEXT T2
1230 T4=T1+1
1240 FOR T2=T4 TO N1
1250 U(T2)=T(T2)
1260 NEXT T2
1270 T4=T0+1
1280 FOR T2=T4 TO T1
1290 T5=T1+T0+1-T2
1300 U(T2)=T(T5)
1310 NEXT T2
1320 REM                   ******
1330 REM                   STEP 4
1340 REM                   ******
1350 L1=0
1360 FOR T2= 1 TO N
1370 T4=T2+1
1380 T5=U(T2)
1390 T6=U(T4)
1400 L1= L1 + E(T5,T6)
1410 NEXT T2
1420 IF L1>=L THEN 1480
1430 L=L1
1440 FOR T2=2 TO N
1450 T(T2)=U(T2)
1460 NEXT T2
1470 N9=1
1480 NEXT T1
1490 NEXT T0
1500 IF N9>0 THEN 1130
1510 REM                   ******
1520 REM                   STEP 5
1530 REM                   ******
1540 PRINT "TOUR FOUND"
1550 FOR T0 = 1 TO N
1560 PRINT T(T0);
1570 NEXT T0
1580 PRINT T(N1)
1590 PRINT "THIS HAS LENGTH ";L
1600 STOP
1610 PRINT "TOO MANY NODES,  ALTER
     DIMENSIONS OF PROGRAM"
1620 STOP
1630 PRINT "TOO FEW NODES,  TRY AGAIN"
1640 GOTO 290
1650 PRINT "TOO MANY ARCS,  ALTER
     DIMENSIONS OF PROGRAM"
1660 STOP
1670 PRINT "TOO FEW ARCS,  TRY AGAIN"
1680 GOTO 290
```

6.2.6 An exact method (Little's algorithm)

An iterative method for the travelling salesman problem was developed by Little, Murty, Sweeney and Karel [28] and is exact, in that it will always find the optimal salesman tour. However, it may well prove very expensive computationally, although it is generally reasonably quick. The method is based on a branch-and-bound approach, which builds up a tree of nodes corresponding to particular subsets of the set of all possible tours. Associated with these nodes there are lower bounds which are less than or equal to the shortest possible salesman tour in the corresponding subset. This subset is then further divided, into two comple-mentary subsets, and two new lower bounds found. After repeated division, the subset corresponding to a node consists of just one tour, and the length of this tour is the bound associated with the node.

There are very many ways that these sets and their nodes could be successively created; it is desirable that the number which are needed before the optimum is found be kept as small as possible, since each node requires a certain amount of calculation to find the bound. So the iterative method 'decides' which nodes to explore (by sub-division of their sets) according to the bounds associated with the nodes. The rules for reaching this decision, and for creating and dividing the sets of possible tours for each node, will be described below.

The algorithm starts with the matrix of distances between nodes, D. (This may be the matrix of shortest distances, or the matrix of arc lengths.) This is then transformed to give a 'reduced matrix' D'. D' is derived from D by:

(a) setting $d(i,i) = \infty$ for all i;
(b) subtracting from each element of D, the smallest element in the corres-ponding row, and when this has been done, subtracting from each resulting element the smallest element in the corresponding column;

Thus a matrix D of the form:

$$D = \begin{bmatrix} 0 & 14 & 13 & 17 \\ 10 & 0 & 12 & 19 \\ 9 & 11 & 0 & 8 \\ 15 & 14 & 13 & 0 \end{bmatrix}$$

becomes successively

$$\begin{bmatrix} \infty & 14 & 13 & 17 \\ 10 & \infty & 12 & 19 \\ 9 & 11 & \infty & 8 \\ 15 & 14 & 13 & \infty \end{bmatrix}$$

$$\begin{bmatrix} \infty & 1 & 0 & 4 \\ 0 & \infty & 2 & 9 \\ 1 & 3 & \infty & 0 \\ 2 & 1 & 0 & \infty \end{bmatrix} \quad \begin{matrix} (13 \text{ subtracted}) \\ (10 \text{ subtracted}) \\ (\ 8 \text{ subtracted}) \\ (13 \text{ subtracted}) \end{matrix}$$

$$\begin{bmatrix} \infty & 0 & 0 & 4 \\ 0 & \infty & 2 & 9 \\ 1 & 2 & \infty & 0 \\ 2 & 0 & 0 & \infty \end{bmatrix} = D'$$

(where 1 has been subtracted from the second column)

Part (a) is used to forbid loops from one town to itself; part (b) is a consequence of the (obvious) result that to get to some town from town i, one must travel at least the shortest distance from town i to any other town. Since this distance is essential, one may set it to one side by subtracting it from all the entries in the same row (while recording it as a necessary part of the tour). Similarly, as there must be a town before town i, one can subtract the smallest distance in the corresponding column to make the least entry in this equal to zero. No salesman tour can be shorter than the sum of all the terms which have been subtracted. D' is such that there must be at least one zero element in each row and each column; the identity of the optimum tour in D' is the same as the optimum tour in D.

In some cases, an optimum tour can be found in D' by inspection, using zero entries only. Such a tour exists in the example above: $(1 - 3 - 4 - 2 - 1)$. However, this is unusual, as the example below shows (D is slightly modified from the example given earlier):

$$D = \begin{bmatrix} 0 & 14 & 13 & 17 \\ 10 & 0 & 12 & 19 \\ 9 & 11 & 0 & 8 \\ 15 & 24 & 13 & 0 \end{bmatrix}$$

yields $\quad D' = \begin{bmatrix} \infty & 0 & 0 & 4 \\ 0 & \infty & 2 & 9 \\ 1 & 2 & \infty & 0 \\ 2 & 10 & 0 & \infty \end{bmatrix}$

Attempts to build a tour using the zeroes fail; one could start $1 - 2$, but the only zero in the second row gives the next link as $2 - 1$; or one could start $1 - 3 - 4$, but the next link would be $4 - 3$. So the best salesman tour must use some link which has a non-zero reduced entry. Rather than use a method which tries out the effect of including each one of these in turn, a tour is gradually built up using the zero entries of D', only including the non-zero entries when this is necessary.

The first node corresponds to all possible tours of the network, and the bound for this node is the sum of all the elements subtracted from D to obtain D'. To branch from this node, one arc (i, j) is selected, and the two nodes derived are those corresponding to all arcs which include this link as part of the tour, and those which exclude it. The arc (i, j) is one which has a zero entry in D', and is selected according to the rule 'choose that arc from the list of potential arcs whose exclusion would have the greatest effect'. This can be determined by examination of each potential arc, (i,j). If this arc is not used in the tour, then the tour must use some other arc from i, and some other arc to j. The effect on the bound on the tour is to increase it by the sum of the smallest entry in the ith row of the matrix (ignoring the entry for (i,j) which is zero) and the smallest entry in the jth column (again ignoring the entry for (i,j)). Selecting the arc (i,j) with the largest such penalty cost (or choosing one arc arbitrarily where there is a choice) leads to the first branch in the branch-and-bound tree. For the node which includes arc (i,j) there will be a distance matrix with the ith row and jth column removed, which can be reduced as before to give a lower bound on the tours which correspond to this set. In this matrix, the entry for (j,i) will be set to ∞ to prevent the tour looping. Corresponding to the node for which the set is of all tours excluding the link (i,j), there will be a distance matrix for which the length of link (i,j) will be set to ∞; this too can be further reduced to give a bound on the set of all tours which correspond to this set.

In the case of the matrix D' last used (where the reduction is 45):

penalty for excluding $(1,2) = 0 + 2 = 2$
penalty for excluding $(1,3) = 0 + 0 = 0$
penalty for excluding $(2,1) = 2 + 1 = 3$
penalty for excluding $(3,4) = 1 + 4 = 5$
penalty for excluding $(4,3) = 2 + 0 = 2$

Thus the next nodes are those which include and exclude $(3,4)$.

(a) Matrix when $(3,4)$ is included:

$$\begin{bmatrix} \infty & 0 & 0 \\ 0 & \infty & 2 \\ 2 & 10 & \infty \end{bmatrix}$$

which reduces by 2 to

$$\begin{bmatrix} \infty & 0 & 0 \\ 0 & \infty & 2 \\ 0 & 8 & \infty \end{bmatrix}$$

(b) Matrix when $(3,4)$ is excluded:

$$\begin{bmatrix} \infty & 0 & 0 & 4 \\ 0 & \infty & 2 & 9 \\ 1 & 2 & \infty & \infty \\ 2 & 10 & 0 & \infty \end{bmatrix}$$

which reduces by 5 to

$$\begin{bmatrix} \infty & 0 & 0 & 0 \\ 0 & \infty & 2 & 5 \\ 0 & 1 & \infty & \infty \\ 2 & 10 & 0 & \infty \end{bmatrix}$$

There will now be two nodes from which no branches have been made. That with the smaller lower bound is selected, and the same procedure is repeated, splitting its set into two subsets, which include/exclude a particular arc, chosen according to the penalty of not using it. This will create two new nodes, each with a lower bound, and all the bounds on the nodes will be scanned to find the one whose bound is least. The set associated with this will then be split, and the corresponding bounds calculated in the same way as for the first branch. One important difference will occur, however; in addition to forbidding the loop of two towns, by setting the return link to infinite length, other return loops will be forbidden, since a new link is likely to extend a partial tour.

The algorithm eventually finds a set which only contains one tour, for which the bound is less than or equal to the bounds on every other node. This will be an optimal salesman tour. The fact that it is a salesman tour is obvious, since it will have been built up using the rules for branching, to create sets which either include or exclude certain links. It is optimal because no other node can give rise to a tour with a shorter total distance; every time that a branching is made, the bounds on the tours are increased, or kept constant, but never reduced. So, the fact that no other node can give rise to a shorter tour ensures optimality. However, it is possible that there are other tours of the same length. This will happen if there are nodes whose bound equals the length of the tour which has been found, and which have branchings whose bounds do not increase until a tour is found.

It is usually convenient to summarise the successive stages of branching and evaluation of nodes in a diagram, as in Fig. 6.13. Here, the node at the top

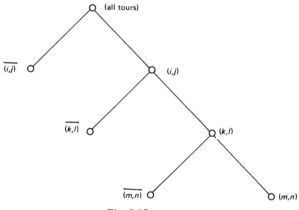

Fig. 6.13.

represents the set of all tours, whose lower bound is the amount by which the original matrix D was reduced to yield D'. This gives rise to two nodes, the one on the left excluding the link (i,j) (signified by $(\overline{i,j})$) and that one the right including this link. The descendants of each of these nodes include/exclude additional links, so the node labelled $(\overline{m,n})$ in the diagram corresponds to the set of tours which include links (i,j) and (k,l) but exclude (m,n).

In Fig. 6.14 the example which was started above is continued, with its corresponding matrices and bounds.

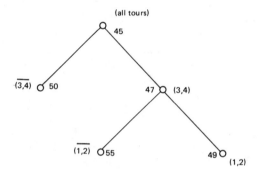

Fig. 6.14 – The node (1,2) corresponds to the tour 1-2-3-4-1.

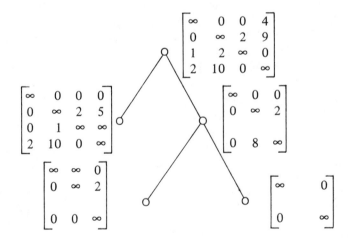

6.2.7 Implementation
This is possibly the most complex algorithm to be considered in this book. There are several points of difficulty, which are interrelated. The first is the way that the tree, and its constituent nodes, are stored and referred to. Then there is the problem of searching through the tree. And then there is the problem of finding which links have been used in, or excluded from, the set of tours to which each

node corresponds. Together, these impose conditions on the amount of data which has to be stored: as a minimum, for each node, the program will require:

(a) the type of node, whether it corresponds to the inclusion or exclusion of a link
(b) the link being included/excluded
(c) the bound on the tours corresponding to this node
(d) the preceding node in the tree
(e) whether or not this node has descendants
(f) whether the node corresponds to a tour.

For searching through the tree, (c) and (e) are essential. Item (e) determines whether the node needs to be examined, and (c) is needed for comparison of different nodes. In a manual search, only the nodes which have no descendants will be considered, and these can be readily identified from the tree diagram − in the computer program, they will need to be easily recognised as well. The last aspect of the node can be checked once a node has been selected, and if a tour has been found, the problem is solved. If a node does not correspond to a tour, then the tree can be explored to its first node by using the preceding node, then the preceding node of this, and so on. As this exploration progresses, the first and second items will yield the links which are to be excluded and included in the tour associated with node. Given these, the distance matrix can be calculated, and the bounds on tours calculated from it.

In practice, (b) above needs to be stored as an ordered pair of nodes, and (a) may be combined with (f). This is possible, since a tour always corresponds to a node which includes a link.

These items provide a part of the necessary structure of the program; they must then be fitted into an overall scheme for referring to a series of nodes, and moving from one to another. There are several ways that this can be done, but the simplest is almost certainly to use an array to hold nodes as they are created. Then reference to any node can be to its position in the array. In the PASCAL program, all the nodes are stored in records, and the tree is held in an array of such records, although a series of arrays could be used instead.

Once a node has been selected to be branched from, the program must provide a means for calculating the link to be included/excluded, and the bounds associated with each possibility. There must also be a test for the inclusion of a link leading to the completion of a tour. So that these can be done, there must be a means of calculating the appropriate distance matrix for the node, together with the links which have been excluded and included, so that not all the rows and columns are examined. This requires a means of finding the effect of reducing a given matrix, and the penalty of excluding links. This can be effected by a routine which reduces a matrix, or some rows and columns of a matrix, and then finds the second smallest entries for each row and column of it. The link to be used next can then be found by selecting the largest penalty of exclusion, using the zeroes of the reduced matrix.

Identification of rows and columns to be excluded from the distance matrix used at a given node means searching through the tree until the first node is reached, and forbidding each row which is the start of an included link, and each column which is the end of such a link. At the same time, the matrix has to be modified by setting forbidden links to infinity. Some of these links will be the excluded links, the others will be the reverse links of paths formed by several included links.

When the node which is being branched from is the last to have been examined, then there is no need to recreate the distance matrix from scratch. That matrix which was last used can be retained, with the minor changes to it which correspond to the inclusion of a new link.

The PASCAL program uses two distance matrices throughout, one being the original which was read in from the user, and one being derived from this to correspond to a given node. The first node, corresponding to 'all tours', is found and created. The program then enters a loop, finding the node with the smallest bound (and, if there are several, the most recently created one. This tends to bias the search towards those nodes which correspond to almost complete tours, and to the last node to be created). Then, this node is branched from. The distance matrix is set up, and so are the vectors which determine which rows and columns have been excluded. These are Boolean vectors, true for rows (columns) which can be considered, false otherwise. Then the next two nodes can be created, and their bounds established, on the basis of the information provided by the reduction of the matrix. Creation of the 'included' node automatically provides the link which will be included by the descendant of this node, and, if there are only two rows eligible for consideration in the matrix for this node, the node can be identified as a tour. The loop ends when the node being examined has been found to be a tour.

Identifying the tour, given the node, is straightforward. The tree is searched, from the final node, backwards, adding to a vector of node-numbers for the original network which stores the tour.

```
program prog6p4(input,output);
  const maxnodes = 30;
         mxnodesp= 31;
         maxarcs = 80;
         treesize = 201;
         inf= 9999;
  type  dismat = array[1..maxnodes,
##      1..maxnodes] of integer;
  ar     = array[1..maxarcs] of integer;
  intarr = array[1..maxnodes] of integer;
         inset = 0..maxnodes;
         nodekind = (firstnode,excluded,
##       included,isatour);
  logarr = array[1..maxnodes] of boolean;
         noderec = record
                     nodetype : nodekind;
```

```
       from.tu.bound.predecessor : integer;
                          branchedfrom : boolean
                 end;
  var    dis.dnext : dismat;
         i.j.d : ar;
         arc.arcs.t1.t2.noofnodes.last.
n.t0.b.theta.ls.ms.K.minbd.l : integer;
         rw.cl.alltrue : logarr;
         besttour : intarr;
         nd : array[1..treesize] of noderec;
         tour : boolean;

  procedure readnet;
  var t8 : integer;
  begin

     write('How many nodes. how many arcs?');
        readln(n.arcs);
           writeln('Enter arcs with their start
##     node finish node length');
           for t8 := 1 to arcs do
           repeat
              write('Arc number '.t8:3.' ');
              readln(i[t8].j[t8].d[t8]);
     if ((i[t8]>n)or(j[t8]>n)) then
writeln('Node number too high; try again');
     if ((i[t8]<=0)or(j[t8]<=0)) then
writeln('Node number too low; try again');
     if (i[t8]=j[t8]) then
           writeln('start and finish must be
##     different; try again ')
     until ((i[t8]<>j[t8])
and(i[t8]in[1..n])and(j[t8]in[1..n]));
     end {readnet} ;
     procedure gettour(lastnode : integer; var
##     thetour : intarr);
     var i.j.K.l : integer;
     begin
     {gettour creates a tour from the tree.
##     given lastnode which is the node which
     corresponds to n-1 links having been set.
##     it creates a fictitious node to
     retain the nth link. and then builds up the
##     tour be searching backwards
     through the tree}
     l := lastnode + 1;
     nd[l].predecessor := lastnode;
     nd[l].nodetype := isatour;
     for i := 1 to n do if (rw[i]) then
##     nd[l].from := i;
     for i := 1 to n do if (cl[i]) then nd[l].tu
##     := i;
     i := 1;
     K := 1;
     thetour[1] := 1;
     while (K<n) do
     begin
       j := 1;
       K := K+1;
     while ((nd[j].nodetype=excluded)or
            (nd[j].from<>i)) do
       j := nd[j].predecessor;
       i := nd[j].tu;
       thetour[K] := i
     end
     end {gettour};
```

```
 procedure findpath(var firstend,lastend :
##    integer; nodeno : integer);
 {this procedure will extend a given link,
##    or part of a path, defined
 by firstend and lastend, to the maximum
##    extent allowed by the branch
 and bound tree, nd, starting from nodeno
##    and working back}
 var last : integer;
 begin
 {extend path backwards}
   last := nd[nodeno].predecessor;
   while (last>1) do
   begin
   if ((nd[last].tu=firstend)
and(nd[last].nodetype<>excluded)) then
     begin
        firstend := nd[last].from;
        last := nodeno
     end;
     last := nd[last].predecessor
   end;
 {extend path forwards}
   last := nd[nodeno].predecessor;
   while (last>1) do
   begin
   if ((nd[last].from=lastend)
and(nd[last].nodetype<>excluded)) then
     begin
        lastend := nd[last].tu;
        last := nodeno
     end;
     last := nd[last].predecessor
   end
 end; {findpath}
 function reduce(var d: dismat; var rw,cl :
##    logarr;
                    var theta,l,m : integer;
 var tour : boolean): integer;
 var row1,row2,col1,col2 : intarr;
     c,i,j,k,r,rwr,clc : integer ;
 begin
 tour := false;
 for i := 1 to n do
 begin
   row1[i] := inf;
   row2[i] := inf;
   col1[i] := inf;
   col2[i] := inf
 end;
 j := 0;
 k := 0;
 for r := 1 to n do
 if (rw[r]) then
 begin
   for c := 1 to n do
   if (cl[c]) then
   begin
     if (d[r,c]<=row1[r]) then
       row1[r] := d[r,c];
       k := k+1
   end;
   j := j + row1[r];
   for c := 1 to n do
     if (cl[c]) then
        d[r,c] := d[r,c] - row1[r];
 end;
```

```
    if (k<=4) then tour := true;

    for c := 1 to n do
    if (cl[c]) then
    begin
      for r := 1 to n do
      if ((rw[r])and(d[r,c]<=col1[c])) then
         col1[c] := d[r,c];
      for r := 1 to n do
      if (rw[r]) then
        d[r,c] := d[r,c] - col1[c];
      j := j + col1[c]
    end;
    for r := 1 to n do
    if (rw[r]) then
    begin
      rwr := inf;
      for c := 1 to n do
      if (cl[c]) then
        begin
        if (d[r,c]<=rwr) then
        begin
          row2[r] := rwr;
          rwr := d[r,c]
        end;
        if ((d[r,c]>rwr)and(d[r,c]<row2[r]))
##      then row2[r] := d[r,c]
      end
    end;
    for c := 1 to n do
    if (cl[c]) then
    begin
      clc := inf;
      for r := 1 to n do
      begin
        if (d[r,c]<=clc) then
        begin
          col2[c] := clc;
          clc := d[r,c]
        end;
        if ((d[r,c]>clc)and(d[r,c]<col2[c]))
##      then col2[c] := d[r,c]
      end
    end;
    l := 1;
    m := 1;
    theta := 0;
    for r := 1 to n do
    if (rw[r]) then
      for c := 1 to n do
      if ((cl[c])and(d[r,
##       c]=0)and(row2[r]+col2[c]>theta)) then
        begin
          theta := row2[r] + col2[c];
          l := r;
          m := c
        end;
    reduce := j;

    end {reduce};

    begin
    { Little's method for the travelling
##    salesman problem
    using a branch-and-bound approach
    variables used
    dis     : matrix of shortest distances for
```

```
##     network
 dnext     : copy of dis used for working
 i,j,d     : vectors used for inputting the
##     network, which is assumed
                to be undirected
 arc       : loop counter
 arcs        number of arcs (<=maxarcs)
 noofnodes:  number of nodes
 last      : last node created
 b         : bound on last node
 theta     : penalty associated with last
##     branching
 ls,ms     : link associated with last branching
 k         : counter for searching through tree
 minbd     : smallest bound on an included node
 rw,cl     : boolean vectors for storing
##     reduced matrix
 besttour: the best tour in the network, as
##     a series of nodes
 nd        : the tree, stored as an aray of
##     nodes with seven properties
         nodetype : included, excluded, etc.
         from,tu  : link associated with the
##     node (to is reserved)
         bound    : the bound on the node
 predecessor: the node above this in the tree
         branchedfrom: boolean to identify
##     used  nodes
 tour      : boolean for finding a tour
 }
 writeln('Branch and bound method for
##     travelling salesman problem');
 readnet;
 for t0 := 1 to n do alltrue[t0] := true;

 for t0 := 1 to n do
 for t1 := 1 to n do
   dis[t0,t1] := inf;
 for arc := 1 to arcs do
 begin
   dis[i[arc],j[arc]] := d[arc];
   dis[j[arc],i[arc]] := d[arc]     {delete
##     this line if directed network used}
 end;
 for t0 := 1 to n do
 for t1 := 1 to n do
 for t2 := 1 to n do
   if (dis[t1,t2]>(dis[t1,t0]+dis[t0,t2]))
##     then dis[t1,t2] := dis[t1,t0]+dis[t0,t2];
 { simplified form of Floyd method }
 for t0 := 1 to n do dis[t0,t0] := inf;
 noofnodes := 1;
 rw := alltrue;
 cl := alltrue;
 dnext := dis;
 last := 1;
 with nd[1] do
 begin
   nodetype := firstnode;
   from := 0;
   tu := 0;
   bound := reduce(dnext,rw,cl,theta,ls,ms,
##     tour);
   predecessor := 0;
   branchedfrom := false
 end;
```

```
{              ******
               STEP 1
               ******
}
k := 1;
last := 1;
repeat
  minbd := inf;
  for t0 := 1 to noofnodes do
  if ((not nd[t0].branchedfrom)
     and(nd[t0].bound<=minbd)) then
  begin
    k := t0;
    minbd := nd[t0].bound
  end;
  nd[k].branchedfrom := true;
  if (k<>last) then
  begin
    dnext := dis;
    rw := alltrue;
    cl := alltrue;
    l := k;
    b := 0;
    t0 := 0;
    t1 := 0;
    t2 := 0;
    repeat
      if (nd[l].nodetype=included) then
      begin
        if (t0=0) then
        begin
          t0 := nd[l].from;
          t1 := nd[l].tu;
          t2 := l
        end; {sets up the link t0,t1 as the
##   last included link}
        rw[nd[l].from] := false;
b := b + dnext[nd[l].from,nd[l].tu];
        cl[nd[l].tu] := false
      end
      else
      if (nd[l].nodetype=excluded) then
        dnext[nd[l].from,nd[l].tu] := inf;
      l := nd[l].predecessor
    until (l<=1);
    if (t0 <> 0) then
    repeat
      if (nd[t2].nodetype=included) then
        findpath(t0,t1,t2);
      if (nd[k].nodetype<>isatour) then
##   dnext[t1,t0] := inf;
{test for the final node of the tree}
      t2 := nd[t2].predecessor
    until (t2<=1);
    b := b + reduce(dnext,rw,cl,theta,ls,ms,
##   tour);
  end;
  noofnodes := noofnodes + 1;
  with nd[noofnodes] do
  begin
    nodetype := excluded;
    from := ls;
    tu := ms;
    bound := minbd + theta;
    predecessor := k;
    branchedfrom := false
  end;
```

```
    noofnodes := noofnodes + 1;
    with nd[noofnodes] do
    begin
       nodetype := included;
       from := ls;
       tu := ms;
       rw[ls] := false;
       cl[ms] := false;
       t0 := ls;
       t1 := ms;
       predecessor := k;
       findpath(t0,t1,noofnodes);
       if (nd[k].nodetype<>isatour) then
##        dnext[t1,t0] := inf;
          bound := minbd + reduce(dnext,rw,cl,
##        theta,ls,ms,tour);
          branchedfrom := false;
          if (tour) then nodetype := isatour
       end;
       last := noofnodes
    until ((nd[k].nodetype=isatour)
          or(noofnodes>=treesize-1));
    if (noofnodes>=treesize-1) then
       writeln('The tree has grown too large for
##        the store.   Increase treesize')
    else
    begin
       gettour(noofnodes,besttour);
       writeln('Optimal tour');
       for t0 := 1 to n do write(besttour[t0]:3,
##        '-');
       writeln('  1');
       writeln('This has length',
##        nd[noofnodes].bound)
    end;
    writeln('Do you want the tree displayed?
##        It occupies',noofnodes,' nodes');
    write('Answer 1 for yes, any other number
##        for no');
    readln(t1);
    if (t1=1) then
    begin
       writeln('The tree is as follows:');
       writeln('Node number From town    To town
##             Bound  Predecessor');
       for t0 := 1 to noofnodes do
       with nd[t0] do writeln(t0:11,from:10,
##          tu:10,bound:10,predecessor:10);
    end
    end.
```

6.2.8 Other algorithms

As has been remarked, there are many alternative algorithms for the travelling
salesman problem. Most of the exact methods (at least for large problems) use
some kind of branch-and-bound approach, and attempt to limit the growth of
the tree wherever possible. It is a subject where there is considerable ongoing
research.

Two methods should be pointed out as being well established, and of
relevance to a book such as this, because of their connection with graph and
network algorithms. The first of these is due to Held and Karp [25, 26], and uses

spanning trees to create bounds for tours. A spanning tree contains $n - 1$ arcs, a tour has one more. In the algorithm, spanning trees are successively modified by assigning penalties to selected nodes, so that the tree becomes progressively closer and closer to being a tour. The second is based on the assignment problem, which was described in Chapter 4. The matrix of distances for the travelling salesman problem is treated as a cost matrix for the assignment problem, and the latter is solved. The resulting assignments are treated as links for the tour. Normally, these links will form a series of disjoint circuits, although sometimes a salesman tour will be found at once. In the former case, a bound for all tours will be found; one of the circuits C is selected, and the nodes in it $(i_1, i_2, \ldots i_p)$ are noted. The optimal tour must include at least one link other than those which are used in C, and so the branching is performed on the basis of forbidding links from the nodes of C to one another in turn. This yields p further bounds, which are explored in the same way again. Eventually the salesman tour is found.

Fuller discussion of these methods can be found in many of the works referred to in the bibliography.

Extensions to the salesman problem

It is apparent that the salesman problem, as described here, is an abstract model of the behaviour of 'real-life' travelling salesmen. There have been several attempts to build a more realistic mathematical model by adding suitable constraints. One of these attempts is concerned with the allocation of several salesman to the tour, to divide the 'territory' fairly between them. Another is concerned with the problems of fitting the tour into a working schedule, so that certain visits are made at particular times of the day. And yet others are concerned with arranging that certain nodes are visited after others. There have been algorithms devised for these problems, which use many of the ideas described already; however, their detail lies outside the scope of this book.

EXERCISES

(6.1) (N, A, D) is an even, symmetric network. Show that there is an Euler tour on this network. (Hint: separate A into a set of directed arcs, and a complementary set of undirected arcs, and show that there are one or more Euler tours (possibly disjoint) on each, which may be spliced together to give an Euler tour on the whole.)

(6.2)   How many bridges need to be repeated in the Konigsberg bridges problem?

(6.3) (N, A, D) is a directed network which has an optimal postman tour. One arc, (i, j), of A is reversed; under what circumstances will

(a)   there be no postman tour on the resulting network?
(b)   the postman tour be shorter?
(c)   the postman tour be longer?

(6.4)   Find an optimal postman tour in the network shown below.

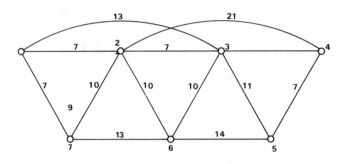

Fig. 6.15.

(6.5)   In the garden city of Urbsisca, the wide grass verges to roads are mown
         in two stages: first by a gang-mower, and then by a self-propelled
         mower. The latter is used to mow the edges of the verges, and (where
         necessary) around the ornamental trees planted in the verges. These
         self-propelled mowers therefore must traverse each verge twice (no trees)
         or three times. Show how the postman problem can be modified for
         finding the shortest tour for the self-propelled mower.

(6.6)   Find an optimum salesman path in the network whose distance matrix is:

$$D = \begin{bmatrix}
\infty & 53 & 20 & 72 & 94 & 47 & \infty \\
53 & \infty & 20 & 98 & 98 & 77 & 94 \\
20 & 20 & \infty & 69 & 59 & 10 & 83 \\
72 & 98 & 69 & \infty & 71 & 90 & 20 \\
94 & 98 & 59 & 71 & \infty & \infty & 66 \\
47 & 77 & 10 & 90 & \infty & \infty & 83 \\
\infty & 94 & 83 & 20 & 66 & 83 & \infty
\end{bmatrix}$$

         Compare this with the path found using an approximate method.

(6.7)   Does the best path found using the 'two-optimal' method depend on
         the way in which the nodes have been numbered?

(6.8)   In a fudge-making factory, several flavours of fudge are mixed in
         sequence on one machine. After each flavour, the machine must be
         cleaned in readiness for the next flavour, and the time spent cleaning
         depends on the two flavours. The production manager wishes to find
         a sequence which minimises the total time spent cleaning the machine
         in each cycle from vanilla to vanilla, making each product once only.
         What is the best sequence, given the following times?

	vanilla	strawberry	next product coffee	chocolate	Devon cream	mint
vanilla	0	100	110	110	60	100
strawberry	150	0	170	150	150	200
coffee	140	170	0	100	150	190
chocolate	140	160	90	0	150	200
Devon cream	50	90	80	90	0	110
mint	170	130	150	150	150	0

(times in minutes)

(6.9) Suppose that the salesman wishes to visit two towns, p and q, together (that is, either go to p and then to q, or go to q and then to p). How can this be included in the salesman problem?

(6.10) Suppose that the salesman wishes to return to the start during his tour of the towns. How can this be included in the salesman problem? Will your solution always lead to a sensible tour, in which the salesman visits several towns before his return?

(6.11) A 'knight's tour' on a chessboard is a sequence of knight's moves which starts in any square of the 8 * 8 board, visit every other square, and return to the starting point. How can such a tour be found, using the travelling salesman algorithm?

(6.12) Is there a tour which visits every vertex of a regular dodecahedron once and once only? This may be represented by the diagram below:

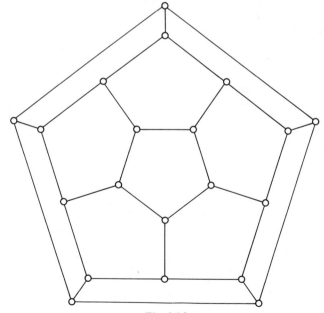

Fig. 6.16.

Further reading

GENERAL

There are numerous books on networks and graph theory, written from a variety of viewpoints, ranging from those of the engineer to those of the pure mathematician. There are three known to the author which are of particular interest to the student of operational research, or to the user of networks who wishes to have an introduction to the practical aspects of the uses of different techniques:

[1]   Christofides, N. (1975), *Graph Theory: An Algorithmic Approach,* Academic Press.
[2]   Deo, N. (1974), *Graph Theory with Applications to Engineering and Computer Science,* Prentice-Hall.
[3]   Minieka, E. (1978), *Optimization Algorithms for Networks and Graphs,* Dekker.

The range of books on programming in PASCAL and BASIC is extremely wide. For the newcomer to PASCAL, the author has found two books particularly useful:

[4]   Moore, L. (1980), *Foundations of Programming with PASCAL,* Ellis Horwood.
[5]   Wilson, I. R. and Addyman, A. M. (1978), *A Practical Introduction to PASCAL,* Macmillan.
[6]   In addition to these, the *PASCAL User Manual and Report* by Kathleen Jensen and Niklaus Wirth (1979), Springer, is invaluable for reference.

For BASIC, the range is even wider; many computer suppliers will provide a manual for their particular dialect. Of the general books, two stand out:

[7]   Alcock, D. (1977), *Illustrating BASIC,* Cambridge.
[7a]   Cope, T. (1981), *Computing Using Basic,* Ellis Horwood.

For a general introduction to computer programming, then without a doubt, one book stands out. This is:

[8] Knuth, D. E. (1968), *The Art of Computer Programming,* Addison–Wesley, Volume 1 (*Fundamental Algorithms*) and Volume 3 (*Sorting and Searching*) being particularly relevant to the material in this book.

BIBLIOGRAPHIES

Chapter 2
[9] Dreyfus, S. E. & Wagner, R. A. (1972), The Steiner Problem in Graphs *Networks,* 1, p. 195.

[10] Gilbert, E. N. & Pollack, H. O. (1968), Steiner Minimal Trees, *Journal of the Society of Industrial and Applied Maths,* 16, p. 1.

[11] Kruskal, J. B. (1956), On the Shortest Spanning Subtree of a Graph and the Travelling Salesman Problem, *Proc. American Mathematical Society,* 7, pp. 48-50.

[12] Prim, R. C. (1957), Shortest Connection Networks and some Generalisations, *Bell System Technical Journal,* 36, pp. 1389-1401.

Chapter 3
[13] Dantzig, G. B. (1967), All Shortest Routes in a Graph, in *Theory of Graphs,* Gordon & Breach, pp. 91-92.

[14] Dijkstra, E. W. (1959), A Note on Two Problems in Connection with Graphs, *Numerische Mathematik,* 1, pp. 269-271.

[15] Floyd, R. W. (1962), Algorithm 97 (Shortest Path), *Communications of the Association for Computing Machinery,* 5, p. 345.

[16] Ford, L. R. (1956), *Network Flow Theory,* Rand Corporation.

[17] Pollack, M. (1961), The kth Best Route Through a Network, *Operations Research,* 9, pp. 578-580.

Chapter 4
[18] Ford, L. R. & Fulkerson, D. R. (1956), Maximal Flow Through a Network, *Canadian Journal of Mathematics,* 8, pp. 399-404.

[19] Ford, L. R. & Fulkerson, D. R. (1962), *Flows in Networks,* Princeton University Press.

Chapter 5
There are numerous books on the subjects of critical path analysis and PERT, ranging from introductory texts for students and managers to manuals for particular computer systems. For general reading beyond the introduction given here, the following are suggested:

[20] Lockyer, K. G. (1978), *Introduction to Critical Path Analysis,* Pitman.

[21] Moder, J. J. & Phillips, C. R. (1970), *Project Management with CPM and PERT,* Van Nostrand.

Chapter 6

[22] Balinski, M. L. (1969), Labelling to Obtain a Maximum Matching, in *Combinatorial Mathematics and its Applications* (Edited by Bose & Dowling), Univ. of North Carolina, pp. 585-602.

[23] Bellmore, M. & Nemhauser, G. L. (1968), The Travelling Salesman Problem – a Survey, *Operations Research,* **16**, pp. 538-558.

[24] Edmonds, J. & Johnson, E. L. (1973), Matching, Euler Tours and the Chinese Postman, *Mathematical Programming,* **5**, p. 88-124.

[25] Held, M. & Karp, R. M. (1970), The Travelling Salesman Problem and Minimum Spanning Trees: Part 1, *Operations Research,* **18**, pp. 1138-1162.

[26] Held, M. & Karp, R. M. (1971), The Travelling Salesman Problem and Minimum Spanning Trees: Part 2, *Mathematical Programming,* **1**, pp. 6-25.

[27] Kwan, M-K. (1960), Graphic Programming Using Odd or Even Points, *Acta Math Sinica,* **10**, pp. 263-266 and *Chinese Mathematics,* **1**, pp. 273-277.

[28] Little, J. D. C. *et al.* (1963), An Algorithm for the Travelling Salesman Problem, *Operations Research,* **11**, pp. 972-989.

Index